CONSERVATION LAWS AND SYMMETRY:
Applications to Economics and Finance

CONSERVATION LAWS AND SYMMETRY:

Applications to Economics and Finance

Edited by

Ryuzo Sato
New York University

and

Rama V. Ramachandran
New York University

The Center for Japan-U.S. Business and Economic Studies
Leonard N. Stern School of Business
New York University

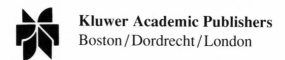

Kluwer Academic Publishers
Boston/Dordrecht/London

Distributors for North America:
Kluwer Academic Publishers
101 Philip Drive
Assinippi Park
Norwell, Massachusetts 02061 USA

Distributors for all other countries:
Kluwer Academic Publishers Group
Distribution Centre
Post Office Box 322
3300 AH Dordrecht, THE NETHERLANDS

Library of Congress Cataloging-in-Publication Data

Conservation laws and symmetry : applications to economics and finance
/ edited by Ryuzo Sato and Rama V. Ramachandran.
     p.   cm.
     ISBN 0-7923-9072-5
     1. Economics, Mathematical.   2. Symmetry (Physics)
3. Conservation laws (Physics)   4. Stochastic processes.
5. Microeconomics.   I. Sato, Ryuzo, 1931-   II.Ramachandran,
Rama V.
HB135.C638   1990
330'.01'51—dc20   90-4190
                                                                CIP

Printed in the United States of America

CONTENTS

Preface

Modern geometric methods combine the intuitiveness of spatial visualization with the rigor of analytical derivation. Classical analysis is shown to provide a foundation for the study of geometry while geometrical ideas lead to analytical concepts of intrinsic beauty. Arching over many subdisciplines of mathematics and branching out in applications to every quantitative science, these methods are, notes the Russian mathematician A.T. Fomenko, in tune with the Renaissance traditions.

Economists and finance theorists are already familiar with some aspects of this synthetic tradition. Bifurcation and catastrophe theories have been used to analyze the instability of economic models. Differential topology provided useful techniques for deriving results in general equilibrium analysis. But they are less aware of the central role that Felix Klein and Sophus Lie gave to group theory in the study of geometrical systems. Lie went on to show that the special methods used in solving differential equations can be classified through the study of the invariance of these equations under a continuous group of transformations. Mathematicians and physicists later recognized the relation between Lie's work on differential equations and symmetry and, combining the visions of Hamilton, Lie, Klein and Noether, embarked on a research program whose vitality is attested by the innumerable books and articles written by them as well as by biologists, chemists and philosophers.

The economic applications of Lie groups were pioneered by Ryuzo Sato. The early results were rigorously and systematically presented in *The Theory of Technical Change and Economic Invariance* by Ryuzo Sato (Academic Press, 1981). In 1987 a conference was organized at the Center for Japan-U.S. Center for Business and

Economic Studies to examine recent researches in this area. We were enthused by the papers presented; while some of the papers continued the line of research begun earlier, others sought out new applications. It was decided to publish a book including not only the papers presented there but a few others that were relevant. All the authors were also encouraged to revise the papers so as to make them as accessible as possible to the non-specialists. Therefore this book is not a conference volume in the conventional sense.

The first chapter by Ryuzo Sato and Rama Ramachandran begins with an examination of the use of symmetry in the study of geometric systems. We argue that the observational implications of symmetry should be considered in economic modelling also. Next we reformulate some mathematical concepts familiar to economists and finance theorists in the language of differential geometry and then introduce additional concepts in an intuitive manner. Finally we use these concepts to examine the Clapham-Pigou and Stigler-Solow controversies and to derive conservation laws (scalar functions that determine an optimal trajectory by the condition that they remain constant along the trajectory) for some well-known dynamic economic models.

The next five chapters that deal with conservation laws are followed by two that examine deterministic and stochastic choice while the final two chapters consider econometric issues.

The second chapter is a reprint of the classic paper in which Paul Samuelson explicitly introduced a conservation law into theoretical economics. It is true that Frank Ramsey implicitly used another conservation law in the derivation of his famous rule for optimal savings rate. It was left to Samuelson to use the analogy of the conservation of energy and derive an explicit conservation law: the constancy of the capital-output ratio in the neo-classical von-Neumann economy.

The next chapter is another paper by Samuelson where he extends and elaborates the model in the earlier chapter. He shows that the output-capital conservation law is derivable from two separate conservation laws: conservation of output measured using its shadow price and the conservation law of capital measured in its shadow price. He also proves similar relations by considering the minimum-time formulation. A very special discrete-time model is also discussed.

Ryuzo Sato seeks, in the article reprinted as chapter three, to derive two conservation laws for models with heterogenous capital. The first conservation law is a pseudo-net-productivity relation which implies that the rate of change in national income is equal to the discount rate multiplied by the utility-value-of-investment. Next, the "modified income" conservation law is confirmed when the economy possesses taste or technical change. The paper ends with a section on the econometric testing.

In the next chapter, Fumitake Mimura and Takayuki Nono discuss a method for deriving conservation laws using the Helmholtz conditions. The latter refers to the necessary and sufficient conditions for a given system of differential equations to be identified with the Euler-Lagrange equations for some Lagrangean. The procedure for the case when $n = 1$ is outlined and then applied to three neo-classical growth models: the Ramsey type model, the Liviatan-Samuelson model and another due to Samuelson. The approach enables one to derive Noether and Non-Noether conservation laws.

Chapter six is by Ryuzo Sato and Shigeru Maeda. One of its purposes is to divide the known conservation laws in continuous dynamic models into five basic types with the help of the Noether theorem. Discrete time models play an important role in dynamic economic analysis and the other objective of the paper is to introduce a new method for analyzing them. This new methodology is discussed in detail. It is shown that there exists several new conservation laws unknown in the continuous case. The appendix to the paper summarizes the various conservation laws in a table.

Thomas Russell provide a differential geometric framework for the analysis of individual choice under uncertainty. The method is used to examine the characterization of choice behavior, single period portfolio theory and asset pricing, and finally continuous time portfolio theory. The chapter seeks to indicate the power of geometric methods as a unifying devise in the study of individual behavior under uncertainty.

In chapter eight, John Boyd III presents a geometric approach to dynamic economic problems that integrates the solution procedure with the economics of the models and reduces reliance on ad hoc techniques. Symmetries explain the form of the solution in terms of the geometry of preferences and technology. Deterministic and stochastic processes with finite or infinite time horizons and constant or variable rates of time preference are treated in a unified manner.

Paul S. Calem examines some issues in the estimation of the rate and bias of technical change and economies of scale. He also examines whether a misspecification can hide the identification problem. Finally he presents formulas that can be used in conjunction with a production function estimation to compute the bias of technical change.

In chapter ten, Thomas Mitchell calls attention to the dearth of econometrically identifiable functional forms for technical change functions and develops means with which the researcher can generate new forms of technical change functions satisfying the appealing properties of transformation groups. These technical change transformations can be embedded in standard econometric models of producer behavior with technical change.

It is now our pleasure to acknowledge our gratitude to a large number of individuals who assisted us with the organization of the symposium that led to this volume and the preparation of the book. We thank all the authors for their co-operation. Two chapters are reprints of journal articles. Chapter two was published in the *Proceedings of the National Academy of Science, Applied Mathematics Science*, vol. 67 (1970), pp. 1477–1479; as reprinted it incorporates minor revisions made by Paul Samuelson. Chapter four was published in

Journal of Econometrics, vol. 30 (1985), pp. 365 -389. We thank the authors and publishers of these papers for copyright permission to reprint them. Jess Benhabib, Gerard Gennatt, Terry Marsh, Hugo Sonnenschein and Shunichi Tsutsui participated in the symposium. Sandra Wren and Lisa Fondo who were with the Center earlier, and Judy Johnson, Ann Barrow, and Kathleen Maloney who are currently with it, provided excellent staff support. Graduate students, Yorum Gelman and Chengping Lian, conscientiously proofread the manuscript. Jim Cozby undertook the arduous responsibility of making the camera-ready copy. Last but not least we must thank Zachary Rolnik of Kluwer Academic Publishers for his patience and encouragement.

Ryuzo Sato

Rama Ramachandran

CONTRIBUTORS

John H. Boyd III, Department of Economics, University of Rochester, Rochester NY 14627, U.S.A.

Paul S. Calem, Federal Reserve Bank of Philadelphia, Philadelphia PA 19106, U.S.A.

Shigeru Maeda, Department of Information Science and Intelligent Systems, Faculty of Engineering, Tokushima University, Minami-Josanjima, Tokushima, 770 Japan

Fumitake Mimura, Department of Mathematics, Kyushu Institute ofTechnology, Tobata, Kitakayushu, 804 JAPAN

Thomas A. Mitchell, Department of Economics, Southern Illinois University, Carbondale IL 62901, U.S.A.

Takayuki Nono, Department of Mathematics, Faculty of Engineering, Fukuyama University, Higashi-Muramachi, Fukuyama, 729-02 Japan

Rama Ramachandran, The Center for Japan-U.S. Business and Economics, Leonard N. Stern School of Business, New York University, New York NY 10006, U.S.A.

Thomas Russell, Department of Economics, Santa Clara University, Santa Clara CA 95053, U.S.A.

Paul A. Samuelson, Department of Economics, Massachussets Institute of Technology, Cambridge MA 02139, and The Center for Japan-U.S. Business and Economic Studies, Leonard N. Stern School of Business, New York University, New York NY 10006, U.S.A.

Ryuzo Sato, The Center for Japan-U.S. Business and Economic Studies, Leonard N. Stern School of Business, New York University, New York NY 10006, U.S.A.

CHAPTER 1

Symmetry: An Overview of Geometric Methods in Economics

Ryuzo Sato
Rama Ramachandran

1. INTRODUCTION

Symmetry is the study of mapping of a state space into itself that leaves a geometric object, generally a set of subspaces defined by an equivalence relation, invariant. Thus, in economics, we can examine whether there exists a transformation to which the indifference curves, subspaces defined by a preference relation, are invariant. However to appreciate the relevance of such a question, it is necessary to have an understanding of the basic principles of geometric spaces. The idea that the quantitative variables of a science are describable by geometric objects and that the laws governing these variables are expressible as geometric relation between the objects,can be traced back to Felix Klein's inaugural address at the Erlanger University in 1872.[1]

Every science identifies, as its field of study, a subset of properties of a given set of objects. In consumer theory, the objects are human beings and the properties are their preferences and purchasing power. In human biology, the properties studied may be the genetic characteristics of the individuals. A demographer has the same objects though he would be interested in their age distribution and sex composition. A mathematician, in contrast, examines the cardinality of the sets, whether the elements are human beings or pebbles.

Further, the level of abstraction in the definition of an object in a science can vary between its theories; particularly relevant for the study of geometry is the changes in the conception of space. Felix Klein lived at a time when the universality of Euclidean geometry was being questioned. The necessity for a reformulation of geometry to incorporate the new developments was obvious. Klein, influenced by the French mathematician Camille Jordan, used groups to classify geometries. Consider elementary Euclidean geometry that we all studied in school; it considers two types of "equalities" between figures (Yaglom, 1988, pp. 111-24). We know that two figures are *congruent* (equal in one sense) if there exists an isometry (mapping that preserves distances between points) from one figure to another. But in another sense of equality, figures are equal if they are *similar* (mapping that preserves ratio between segments). Many theorems in school geometry require only similarity and we can think of a geometry which strictly identifies similar objects.

Notice that a similarity mapping or transformation satisfies the following relations:

1. *Identity:* There is a transformation, T_0 , that sends a figure to itself (see figure 1).

2. *Inverse:* If there is a transformation, T_1, which sends F_1 to F_2, then there is an inverse transformation T_1^{-1} which sends F_2 to F_1.

3. *Closure:* If there is a transformation, T_1, from F_1 to F_2 and another, T_2, from F_2 to F_3, then the set of transformations contains the "product" transformation $T_1 T_2$ from F_1 to F_3.

4. *Associativity:* The product of two transformations with a third equals the product of the first with the other two (taken in the same order); $(T_1 T_2) T_3 = T_1 (T_2 T_3)$.

A set of transformations that satisfy the above conditions form a *group* of transformations.

We can now introduce the concept of distance and define equality in the sense of congruence. The set of congruence transformations form a group as it satisfies the four relations stated above. A geometry based on congruence will be different from the one based on similarity

Figure 1

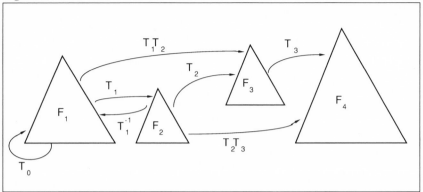

transformations: every congruent transformation is a similarity transformation but the converse is not true.

This is the crux of the idea that Klein exploited to classify geometries. As generally stated, the Erlanger Program defined geometry as the study of the properties of figures invariant to a given group of transformations.[2] Klein's perspective shed light on the theory of relativity in physics. Newtonian physics postulated an absolute space in which all physical bodies are embedded. But its laws of motion cannot distinguish between a coordinate system which was at rest relative to the absolute space and one which was moving at a uniform velocity; in other words, if we sat in an airplane which was flying at a constant speed, then we would not observe anything *within the plane* to distinguish it from one at rest. The laws of Newtonian mechanics are invariant to a Galilean transformation (corresponding here to a relative motion of coordinate systems at constant velocity).

In contrast, consider electromagnetic waves which travel through space with a constant velocity c. If the earth was travelling with a velocity v, in the same direction as a ray of light, then the speed of light as measured on earth should read $c - v$ (just as a plane overtaking yours should appear to be travelling slower than its speed relative to the earth). The famous Michelson-Morley experiment disproved this hypothesis when it showed that the measured speed of light was the same in all directions. The special theory of relativity explained the experimental result by replacing the Euclidean assumptions with a space-time continuum and Lorentz transformation. Newtonian mechanics became a special case for systems that move with velocities substantially below that of light.

The justification for rejecting Galilean transformations (for a wider class of transformations) was that it had consequences which could not be verified experimentally. While Klein originally looked upon the group of transformations as a taxonomic tool, Einstein and philosophers of science like Reichenbach (1960) integrated it into the methodological foundations of physics[3].

In consumer economics, the problem of utility measurement was widely discussed. In the intellectual ferment of the 1930s, marginal utility was dethroned from the preeminent position it had held in consumer theory. Instead of assuming that we know the consumer's utility surface, the new theory assumed that we know only his indifference curves or level sets. Hicks (1946, p. 17) noted that this approach had wide methodological significance as the indifference curves conveyed less information than utility surfaces. Though each set of indifference curves could be thought of as the contour lines of a utility surface, the numbers were arbitrary as long as they preserved the order and so the utility surface could be replaced by a monotonic transformation of itself.

Starting from a given scale of preferences, attention was confined to the properties of the utility function which were invariant to the monotonic transformation. All theorems with observable consequences like the equilibrium and stability conditions could be derived from these properties. Alchian (1953) emphasized sets of transformations in a lucid presentation of the meaning of utility measurement. He noted that measurement in the broadest sense was the assignment of numbers to entities. He distinguished between the purposes of measurement, the process of assigning numerical values to entities and the degree of arbitrariness of such assignments.

Alchian noted the differences between transformations unique up to an addictive or multiplicative constant, general linear transformations, and monotone transformations. In neo-classical theory, differences in the utility levels of various bundles of goods had to be ranked and the numerical assignments had to be unique up to linear transformations. In the ordinal approach, only the basket of goods needed be ordered and utility had to be unique up to monotonic transformations only. Contemporaneous with the revolution in demand theory, criticisms of interpersonal comparisons of utility resulted in the use of the Pareto-principle rather than sum-ranking in welfare economics. The ordinal approach which assumed less information about individual's preferences, became standard in economics. In these discussions, the non-uniqueness in the assignment was emphasized but the idea that the set of transformations form a group was not utilized. The ordinal utility can be identified with the symmetry group of the transformations that leaves the equivalence classes of the preference relation, invariant.

At first sight it would look that mechanics concerned itself with the transformation of coordinate axes while economists were concerned with the transformation of (utility) surfaces. Coordinates were assigned to a point when it was associated with an ordered n-tuple of real numbers $(x^1,\ldots, x^n)$ with distinct points associated with distinct n-tuples. As Reichenbach (1960, p. 90) noted, it was not inherent in the nature of reality that space should be described by coordinates; it was a subjective assignment whose empirical implications had to be examined. Further, all coordinate transformation could be treated as groups of manifold transformation (Friedman, 1983, p. 56). The distinction is more apparent than real.

Consider the effects of technical progress on surfaces defined by a linear homogeneous production function with Hicks-neutral technical progress. The production function can be written as $Y = A(t) f (K, L)$ $= F_t (K, L)$ or $Y = f [A(t) K, A(t)L]$. In the first formulation, the input space is not affected by technical progress but the production surface shifts over time. In the second, the input axes based on efficiency units are transformed by technical progress but the surfaces represented by the functional relation $f ( )$ is not affected. These considerations suggest that technical progress can be viewed as a transformation of the production surface in a given input space or as a transformation of the input space.

As early as 1920s, Clapham (1922) objected to the separation of changes in the efficiency resulting from differences in size from that arising from inventions. The debate with Pigou (1922) that ensued was widely studied in economics. Thirty-five years later, Solow (1957) published his famous article attributing 87.5 percent of the increase in per capita productivity in the United States between 1909 and 1949, to technical changes defined as shifts in production function. Since then innumerable articles and books have sought to explain productivity increases. In one such attempt, Stigler (1961) used US and British data to allocate a sizable portion of productivity growth to returns to scale. Stigler used cross-section data across the countries to get around the conceptual problems of confining such analysis to one economy. Solow (1961, p. 67), in his comments, pointed out that Stigler's approach would not solve the econometric problems and argued that separating economies of scale from technical change "is an econometric puzzle worthy of everybody's talents."

Most econometric studies use Hicks-neutral technical progress. Sato and Ramachandran (1974) examined a production process moving along an expansion path, in input space, at a given rate. Using a technique developed by Zellner and Revankar (1969), we derived a differential equation relating growth in output and in inputs under two alternate assumptions about the production process: first, linear homogeneous production function with Hicks-neutral technical progress, and second, a homothetic production function. We showed that there was a one-to-one mapping from one differential equation to the other. The informational implication was that, from the analysis of data alone, we could not distinguish between the two models. The paper did not explicitly use group-theoretic methods and could not be extended to other forms of technical progress. That step was taken in Sato (1975,1981) when he used Lie groups to examine all known forms of technical progress functions and derive the corresponding form of scale effects.

The Norwegian mathematician and collaborator of Felix Klein,

Sophus Lie, took up the notion of continuous transformation implicit in concepts like the Galilean and Lorentz groups and applied it to the classification, not of geometries, but of differential equations. He showed that the plethora of special methods used in solving differential equations hid the connection between various types of solvable equations, namely the invariance of the integrals to certain types of transformations. He also showed that the invariance could be established without first solving the equation; this in turn, help in determining whether an equation had a solution.

Interest in Lie groups of differential equations waxed and waned in the twentieth century. By 1930s, as Yaglom (1988, pp. 107–108) points out, this aspect of Lie's work no longer elicited much enthusiasm.[4] With the development of computers, solvability in quadratures lost its previous importance. But by 1970s, physicists and later mathematicians rediscovered that Lie's theory not only characterized solvability but also symmetry and there came about a resurgence of interest in his theory. We shall outline this theory in the next section and then proceed to use symmetry to analyze the productivity problem and also derive the conservation laws for two economic models.

Group theory cannot pull rabbits out of a hat. It is a useful tool to identify the informational implications of various assumptions and encourage the development of theories that correspond to the parsimonious information that is available. The concern of Pigou (1922, p. 451) that Clapham's proposal cannot be adopted without injury to the *corpus* of economics is understandable. Returning to Sato and Ramachandran (1974) formulation, microeconomics considerations suggest that a competitive firm under increasing returns should not be expanding at a given rate but must seek to increase it indefinitely. One is left with the option of modifying the theory without waiting for an econometric millennium or expeditiously improving our empirical methods. Sixty-five years of debate since the Clapham-Pigou controversy has failed to produce a conclusive development in either direction.

In economics as in many other sciences, geometric methods are used to characterize quantitative variables by geometric objects. But this method also brings out the fact that geometric objects are independent of the coordination adopted; the latter introduces implicitly or explicitly additional assumptions. As noted above, the implications of numerical assignments or coordination were recognized by economists in a number of contexts and solutions specific to each one of them were adopted. But, unlike in physical sciences, there was no comprehensive consideration of the implications of the invariance of geometric objects to groups of transformations.

II. TOOLBOX

Every discipline has its own terminology and analytical tools. In this section, we outline the mathematical structure needed to understand the geometric approach to symmetry. We begin by taking a fresh look at some concepts which are widely used in economics and finance and then proceed to add new concepts.

2.1 Mapping

A *map, mapping* or *function*, $f : X \rightarrow Y$, from a set X to a set Y, is a rule which associates an element $y \in Y$ with each element $x \in X$; to indicate which element of Y is associated with one in X, we can write $f : x \rightarrow y$. The set of elements of Y to which the elements of X are mapped is the *image* of X, $\text{im} f = \{ f(x) \in Y \mid x \in X \}$. If $\text{im} f = Y$, then f is *into* map or *surjection*.

However, it is not necessary that a unique element $x \in X$ to be associated with any element $y \in Y$. The symbol $f^{-1}(Y)$ represents the set of elements included in X that are mapped into Y by f, the *preimage* of Y. If $f^{-1}(y)$, the preimage of any element y, consists of a single element, then the map $f : X \rightarrow Y$ is called *one-to-one* or an *injection*. The symbol $f^{-1}(\ )$ is a map only if f is an injection.

If the map is both an injection and a surjection, then it is a *bijection*. Further, if $f : X \rightarrow Y$ and $g : Y \rightarrow Z$, then $g \bullet f : X \rightarrow Z$. By convention, the right-hand most mapping is done first and then the one to the left; the same principle can be applied to the composition of more than two maps. If $z \in Z$ is the element to which $x \in X$ is mapped by $g \cdot f$, then we can write $g \cdot f : x \rightarrow z$. For a bijective function f, $f^{-1} \cdot f$ maps an element of x into itself; it is the identity map.

Notice that we have not said anything about the nature of the elements of sets X, Y, or Z. The readers are, of course, familiar with functions between sets whose elements are real numbers. Even if the elements of X and Y are not numbers, we may associate an ordered n-tuple of real numbers with each element. Thus, in Euclidean geometry, the elements of space are associated with a 3-tuple of ordered real numbers; in special theory of relativity, space-time is represented by a 4-tuple. This is the process of coordination referred to in section I.

We may think of a map from X to Y either as an abstract map from the elements of X to the elements of Y or from the coordinates of $x \in X$ to the coordinates of $y \in Y$; the first is a coordinate free definition while the second is not.

2.2 Charts and Manifolds

Every point on the surface of the globe, except the North and South Poles and the Greenwich meridian can be represented by two numbers, the longitude and latitude, that can be taken to be the coordinates of the point. Hence parts of the globe are "locally" like the R^2 plane as one recognizes by looking at the pages of an atlas. The poles are excluded as every longitudinal curve passes through the poles and the meridian because it is both the 0 degree and 360 degree East; in either case the uniqueness condition is infringed. Theoretically one can take two other points on the globe, call them East and West poles, and map every point other than the two new poles and a new meridian (but including the North and South Poles and the Greenwich meridian) using a new system of "longitudes and latitudes." The two systems will together give a one-to-one mapping of every point of the globe to R^2. A stereographic map is more efficient in that it maps the whole surface of the globe, except a pole, into R^2.

A pair consisting of an open neighborhood of $x \in X$ and an injective map to an open subset of R^n is called a *chart*. The set X is a *manifold* if each element belong to a chart.[5] The set of charts that cover a manifold is appropriately called an *atlas*. Note that at least two charts are needed to cover the surface of earth.

A point on the manifold may belong to two charts. Thus in the two systems of longitude and latitude proposed earlier, every point on the globe other than the four poles and the two meridians, belong to both

Figure 2

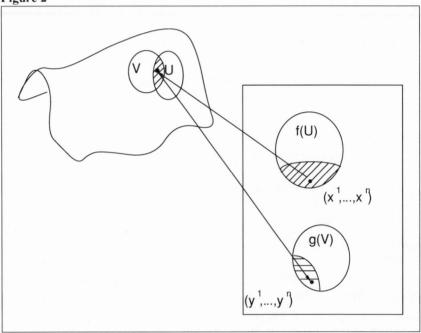

sets. Let f be the function that maps a point to its first set of coordinates $(x^1, \ldots, x^n)$ and g the function that maps it to the second set of coordinates $(y^1, \ldots, y^n)$. On a point where the charts overlap, one can use the principle of composition of maps to obtain a functional relation between one set of coordinates and another. Thus $g \bullet f^{-1}$

defines y^i, $i = 1, \ldots, n$ as a function of $(x^1, \ldots, x^n)$; this is called coordinate transformation. A set of charts of a manifold such that $g \cdot f^{-1}$ is C^k for every point where it is defined, is a C^k manifold. If $k \geq 1$, then the manifold is a differentiable manifold.

If a metric is defined on a manifold, then the open sets can be expressed in terms of the standard $\varepsilon - \delta$ definition. But metric is an additional structure and manifolds can be defined without it. Misner, Thorne and Wheeler (1973, p. 8) has an interesting example. Households with telephone can be identified by the telephone numbers which also identify the locality where the households are located. But one cannot determine from telephone numbers, how many meters away any two of them are.

Three points should be noted. First, surface of earth can be visualized without latitudes and longitudes; coordinates are not essential to the reality of the geometric objects. Second, we instinctively think of the earth's surface as part of a three dimensional space; the existence of such a space is not an integral part of the definition of the manifold given earlier. *An idealized ant crawling on the earth can visualize the surface without seeing the third dimension.* Third, the centrality of the concept of manifolds is brought out by the abstract definition of Lie groups as a differential manifold with group properties.[6]

2.3 Curves and Functions

A demand curve can be thought of as a continuous line, each point of which corresponds to one value of the quantity demanded. This idea can be generalized. A *curve* is a differential mapping from an open interval of a real line to a manifold M. If the points on the open interval are represented by the real number λ, then the curve is *parameterized* by λ and each point of the curve $C(\lambda)$ is an image point of λ. If the differential manifold is coordinated, then the coordinates are differential functions of λ. Again notice that this definition is independent of a metric so that distance along the curve is not defined.

Figure 3

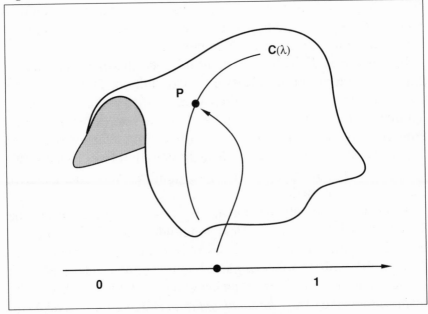

A function is a rule for assigning real numbers to the points of a manifold *M*. If the differential manifold has coordinates, then the function, expressed in terms of the coordinates, takes the familiar form

$$y = f(x^1, \ldots, x^n) \tag{1}$$

The value of this function along a curve $C(\lambda)$ is given by

$$y = f(x^1(\lambda), \ldots, x^n(\lambda)) \equiv g(\lambda) \tag{2}$$

2.4 Vectors and Tangents

In economics we speak of the surfaces generated by utility or production functions. It is easy to visualize a plane tangent to the surface at a point (figure 4). From P, the point of tangency, we can draw a vector on the tangent plane. For any vector PQ on the tangent plane, it is possible to draw a curve, lying on the surface, that has PQ as a tangent at P. Reversing the idea of a vector as a directed line ("arrow"), we can think of every vector as a tangent to some curve.

The last approach is very useful in making generalizations essential to differential geometric approach. The intuitive approach of drawing vectors in a tangential plane as in figure 4 assumes that the manifold

Figure 4

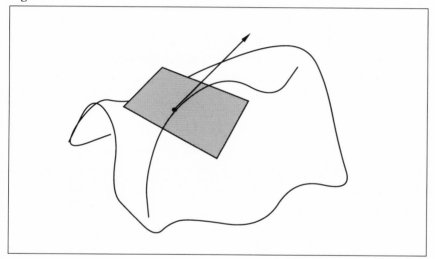

is in a space of higher dimension. As already noted, the existence of a higher dimensional space is not intrinsic to the definition of the manifold. What we need is a definition of a vector which depends only on one point ("local") and not two ("bilocal"); after all, our ant does not know the existence of any point outside the earth's surface. The idea of thinking about a vector as a tangent to some curve on the manifold provides exactly the needed local concept.

Suppose we want to define a tangent vector to the curve $C(\lambda)$ at the point given by $\lambda = 0$. Now define a differentiable function $g : M \to R$ on the manifold; this function has a value at each point of the curve. Let us assume, for the moment, that the manifold is coordinated. Then we can differentiate the function $g(\lambda)$ (see equation (2)) to get

$$\frac{dg}{d\lambda} = \sum_i \frac{dx^i(\lambda)}{d\lambda} \frac{\partial f}{\partial x^i} \tag{3}$$

Now we abstract the operator $d / d\lambda$ from the function g and the coordinates and call it the *directional derivative* (direction being that of the curve $C(\lambda)$), which applied to any arbitrary differentiable function g satisfies equation (3).[7]

The directional derivatives like $d / d\lambda$ form a vector space which provides some intuition why they can be identified with vectors. Consider two curves through P, $C(\lambda)$ and $C'(\mu)$. As before, defining a function g, form the expression

$$a \frac{dg}{d\lambda} + b \frac{dg}{d\mu} \; .$$

We can now draw another curve through P with parameter θ such that

$$\frac{dg}{d\theta} = a \frac{dg}{d\lambda} + b \frac{dg}{d\mu}$$

showing that the operator $d / d\lambda$ satisfies the vector addition law

$$\frac{d}{d\theta} = a \frac{d}{d\lambda} + b \frac{d}{d\mu} \tag{4}$$

To examine the relationship from another angle, assume that the space is coordinated. Then $\partial g / \partial x^i$ gives the partial derivative of the function in the direction of the coordinate axis e_i. This is true for all functions and we can write $\partial / \partial x^i$ as a directional derivative. The set of n such directional derivatives form the basis of the tangent space TM_P at the point P and we can write

$$\frac{\partial}{\partial \lambda} = \sum v^i \frac{\partial}{\partial x^i} \qquad (5)$$

where $v^i = dx^i (\lambda) / d\lambda$ refers to a particular curve. This is similar to the expression for any vector in terms of basis vectors, $v = \sum v^i e_i$. Hence we consider $\partial / \partial x^i$, $i = 1, ..., n$, as the basis vectors of the tangent space.

Having drawn the tangent space at P, we can draw tangent spaces at all other points on the manifold. The collection of these tangent spaces forms the *tangent space to the manifold*. Vector operations are defined on TM_P or TM_Q (where P and Q are two points on the manifold) but a vector on TM_P cannot be added to one on TM_Q.[8] The logic of defining a tangent space as the collection of tangent spaces at individual points is the following. The tangents at P and Q to a curve PQ on the manifold lie on TM_P and TM_Q respectively. Hence the tangent space TM is the "space" in which the tangents to a curve lie.

Another approach to defining vectors is by the rule of their transformation when coordinate bases change (Dubrovin, Fomenko, Novikov, 1984, pp. 146–147). The components of a tangent vector to a curve $C(\lambda)$ in the coordinate system $(x^1, ..., x^n)$ has the components $(dx^1 / d\lambda, ..., dx^n / d\lambda)$. Suppose we now use a new coordinate system $(z^1, ..., z^n)$; let $x^i = x^i (z^1 (\lambda), ..., z^n (\lambda))$ express the old in terms of the new coordinates. Then by chain rule of differentiation,

$$\frac{dx^i}{d\lambda} = \sum_j \frac{\partial x^i}{\partial z^j} \frac{dz^j}{d\lambda} \qquad (6)$$

In the next section, we can contrast this rule of transformation with that of 1-form.

2.5. 1-Forms

Even if the formulation of the last section may look unusual, vector analysis is widely used in economics and finance. Differential forms are not so familiar but they are among the most useful concepts in differential geometry.

The simplest type of a differential form is the 1-form. It can be defined as an operator or a "machine" which outputs a number when a vector is input into it. We represent 1-form by Greek letters. Inserting vector v into a 1-form produces an output $<a, v>$ or $a(v)$. In addition to being a real valued function, 1-forms are linear machines so that

$$<\alpha, au + bv> = a<\alpha, u> + b <\alpha, v> \tag{7}$$

Also we can add 1-forms

$$<\alpha + \beta, v> = <\alpha, v> + <\beta, v>$$

The 1-forms constitute a vector space TM_p^* which is the *dual* of the tangent space TM_p at P of the manifold.

A familiar example of 1-form is the gradient, though the gradient is presented in elementary calculus as a vector. The relation between the gradient df and the directional derivative $d/d\lambda$ is brought out by the equation (Misner, Throne and Wheeler, 1975, pp. 59–60)

$$<df, \frac{d}{d\lambda}> = df\left(\frac{d}{d\lambda}\right) = \frac{df}{d\lambda}. \tag{8}$$

df is, therefore, a machine for calculating the change in f along any desired vector v. This can be represented by an isoquant diagram where v represents a direction along the input space. df then calculates the number of isoquants pierced by v. In elementary calculus, a particular vector corresponding to the direction that maximizes the increase of the output per unit length of the vector is called the gradient; it is the direction of steepest ascent. Schutz (1980, p. 54) points out the limitation of this definition; it depends on the concept of length which is yet undefined. The definition of 1-form as a linear operator is not associated with any metric.

Figure 5

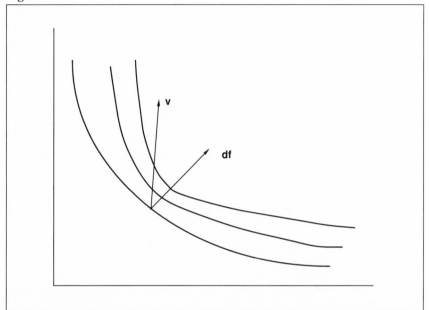

The linear space TM_p^* has the same dimension as the tangent space. Once a basis $(e_1, ..., e_n)$ is chosen for the tangent space, the dual basis is given by the 1-forms e^i or ω^i, $i = 1, ..., n$. Here ω^i is the 1-form that operates on v to produce the ith component of the vector.

$$<\omega^i, v> = \omega^i(v) = v^i.$$

Given any coordinate basis $(x^1, ..., x^n)$, a natural base for the tangent space is $(\partial / \partial x^1, ..., \partial / \partial x^n)$. The base for dual base is $(dx^1, ..., dx^n)$ as

$$<dx^i, \frac{\partial}{\partial x^j}> = \frac{dx^i}{dx^j} = \begin{cases} 1 \text{ if } i = j \\ 0 \text{ if } i \neq j \end{cases} \qquad (9)$$

The components of 1-form df denoted as $f_{,i}$ is given by

$$df = \sum_i f_{,i} \, dx^i \qquad (10)$$

where

$$f_{,i} = <df, \frac{d}{dx^i}> = \frac{\partial f}{\partial x^i} \qquad \text{(see equation (8))}.$$

Hence

$$df = \sum_i \frac{\partial f}{\partial x^i} dx^i$$

where df is the rigorous form of the differential of elementary calculus.

Consider finally the change of coordinates from $(x^1, ..., x^n)$ to $(z^1, ..., z^n)$. Then

$$\frac{\partial f}{\partial z^j} = \sum_i \frac{\partial x^i}{\partial z^j} \frac{\partial f}{\partial x^i} \qquad (11)$$

Compare (11) with (6). Relative to $\partial x^i / \partial z^j$, the expression involving the new coordinates z^k, $k = 1, ..., n$, appear on the left hand side of equation (11) and on the right hand side of (6). In this sense, vectors and 1-forms transform in opposite ways.

2.6. Tensors

Vectors and 1-forms are special cases of tensors. A *Tensor* is defined as a "linear machine" which accepts vectors and 1-forms as inputs and produces a scalar as output.[9]

Figure 6

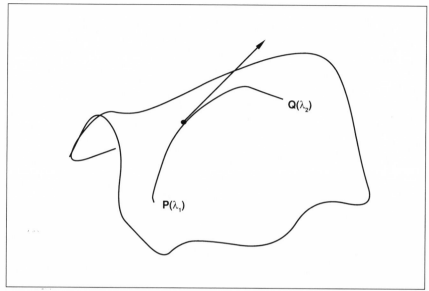

A familiar example is the metric tensor or the distance function which we have assiduously avoided until now. In figure 6, consider the length of the curve $C(\lambda)$ from P to Q. In terms of orthogonal axes, the tangent vector is $(dx^1/d\lambda, dx^2/d\lambda)$. To obtain the length of the curve, we take the square root scalar product of the tangent vector with itself and integrate from $P(\lambda_1)$ to $Q(\lambda_2)$ to obtain

$$\int_{\lambda_2}^{\lambda_2} \sqrt{\left(\frac{dx^1}{d\lambda}\right)^2 + \left(\frac{dx^2}{d\lambda}\right)^2}\ d\lambda$$

But this form, called the Euclidean metric, is very special as can be seen when the distance function is expressed in polar coordinates (r, θ). Then

$x^1 = r \cos \theta,$

$x^2 = r \sin \theta,$

$dx^1 / d\lambda = \cos \theta \, (dr / d\lambda) - r \sin \theta \, (d\theta / d\lambda),$

$dx^2 / d\lambda = \sin \theta \, (dr / d\lambda) + r \cos \theta \, (d\theta / d\lambda).$

Substituting in the expression for length, we obtain

$$\int_{\lambda_1}^{\lambda_2} \sqrt{\left(\frac{dr}{d\lambda}\right)^2 + r^2 \left(\frac{d\theta}{d\lambda}\right)^2} \, d\lambda$$

Instead of unity as the coefficients of the squares under the root sign, we have now 1 and r^2. Further generalizing, the Riemannian metric is defined in terms of a positive definite matrix (g_{ij}) and coordinates $(z^1, \ldots, z^n)$ as

$$l = \int_{\lambda_2}^{\lambda_2} \sqrt{\sum_{i,j} g_{ij}(\lambda) \frac{dz^i}{d\lambda} \frac{dz^j}{d\lambda}} \, d\lambda \tag{12}$$

The matrix (g_{ij}) for the Euclidean and polar coordinates are

$$\begin{bmatrix} 1 & 0 \\ 0 & 1 \end{bmatrix} \text{ and } \begin{bmatrix} 1 & 0 \\ 0 & r^2 \end{bmatrix}.$$

If we think of the process as inputting the same vector twice and obtaining a scalar, then the Riemannian metric is a $\begin{bmatrix} 0 \\ 2 \end{bmatrix}$ tensor. The upper index indicates that no 1-form is accepted by the linear machine. In general, $\begin{bmatrix} 0 \\ 2 \end{bmatrix}$ tensors can accept two separate vectors and generate a number as in the taking of scalar product of two vectors.

In the last section, we defined a 1-form as a machine which accepts a vector and generates a number. If we define vectors as strictly column vectors, then in terms of vector multiplication, a "row vector" accepts a column vector and generates a scalar; hence the row vector can be visualized as a 1-form. Taking the argument one step further, consider a $n \times n$ matrix. By post-multiplying by a column vector and pre-multiplying by a row vector, this matrix generates a number. Hence the matrix can be considered to be $\begin{bmatrix} 1 \\ 1 \end{bmatrix}$ tensor (Schutz, 1980, p. 58).

This leads us to tensor products. The product of a $\begin{bmatrix} 0 \\ 1 \end{bmatrix}$ tensor and a $\begin{bmatrix} 1 \\ 0 \end{bmatrix}$ tensor can be thought of as a linear machine that accepts one vector and one 1-form to generate a scalar. If e_i, $i = 1, \ldots, n$ be the basis of vector space and e^j, $j = 1, \ldots, n$ that of 1-forms then the basis elements of the tensor product are written as $e_i \otimes e^j$. Here i and j takes all the n values independently leading to n^2 basis elements for the tensor. These elements could be written as n^2-tuple with one entry equal to one and all others equal to zero, just as we use an n-tuple to represent a vector (Dubrovin, Fomenko, and Novikov, 1984, p. 154). The notation given above is useful and universally adopted.

In general, if $(e_1, \ldots, e_n)$ are the basis of vectors and $(e^1, \ldots, e^n)$ that of 1-forms, then the $\begin{bmatrix} p \\ q \end{bmatrix}$ tensor can be written as

$$\mathsf{T} = \sum_{i,j} \mathsf{T}^{i_1 \ldots i_p}_{j_1 \ldots j_q} e_{i_1} \otimes \ldots \otimes e_{i_p} \otimes e^{j_1} \otimes \ldots \otimes e^{j_q} \qquad (13)$$

where $\mathsf{T}^{i_1 \ldots i_p}_{j_1 \ldots j_q}$ form a set of numbers for each point in space and are obtained by inputting the basis vectors and 1-forms into the linear machine. The i's and j's take all n values independently leading to n^{p+q} bases. The tensor products of $\begin{bmatrix} p \\ q \end{bmatrix}$ and $\begin{bmatrix} r \\ s \end{bmatrix}$ tensors is a $\begin{bmatrix} p+r \\ q+s \end{bmatrix}$ tensor.

The 1-forms are special cases of differential forms; differential forms, in turn, can be thought of as skew-symmetric tensors of type $\begin{bmatrix} 0 \\ k \end{bmatrix}$ where $k < n$. A tensor is skew-symmetric if $T_\sigma(i_1, \ldots, i_k) = \text{sign } \sigma$

$T_{i_1, \cdots, i_k}$ where σ is a permutation of indices and has a sign $+1$ or -1 according as σ is even or odd (Dubrovin, Fomenko and Novikov, 1984, pp. 163–164). In the case of a two-form

$$\sum_{i,j} T_{ij} \, e^i \otimes e^j = \sum_{i \le j} T_{ij} \, e^i \otimes e^j + \sum_{i \ge j} T_{ij} \, e^i \otimes e^j$$

$$= \sum_{i < j} T_{ij} \, (e^i \otimes e^j - e^j \otimes e^i)$$

The last inequality arises from the properties of skew symmetry, $T_{ij} = -T_{ji}$ and $T_{ii} = -T_{ii} = 0$. Now we define a wedge product as

$$e^i \wedge e^j = e^i \otimes e^j - e^j \otimes e^i \tag{14}$$

and a $[\begin{smallmatrix} 0 \\ 2 \end{smallmatrix}]$ skew symmetric tensor can be written as

$$\sum_{i<j} T_{ij} \, e^i \wedge e^j = \sum_{i,j} T_{ij} \, e^i \otimes e^j$$

Consider the number of independent bases $e^i \wedge e^j$. We already know that there are n^2 expressions, $e^i \wedge e^j$. But the coefficients of $e^i \wedge e^i$, T_{ii}, are zero and the remaining $n^2 - n$ expressions can be bracketed into $1/2(n^2 - n)$ as in equation (14). In general a $[\begin{smallmatrix} 0 \\ k \end{smallmatrix}]$ skew symmetric tensor in an n space has $n! \, / \, (n{-}k)! \, k!$ bases expressed as wedge products of k basis elements.

2.7 Vector Field, Connections and Covariant Derivatives

We have already considered the construction of TM_p, the tangent space at a point. By definition, the tangent space at a point contains many vectors; in fact, it is the set of vectors tangent to the manifold at P. A *field* of vectors X is an assignment of one tangent vector to each point of the manifold. The tangent vector was written as $\sum v^i (\partial / \partial x^i)$ in a local coordinate chart. If f is a differentiable function and if $\sum v^i (P)(\partial f / \partial x^i)$ is a differentiable function for all f, then X is called a *vector field* (Millman and Parker, 1977, p. 216). In short, it is an assignment of vectors to the points of the manifold that is "smooth" in some sense.

Consider vectors at different points of the manifold. In Euclidean geometry, there is an intrinsic notion of parallelism; if two vectors are at different points in this page, we can ascertain whether they are parallel or not. In contrast, consider two vectors tangential to the earth's surface (considered as a two dimensional manifold). Let one vector be in the tangent space at New York and the other in the tangent space at Chicago. For us who see the surface in a three dimensional Euclidean space, the question whether the vectors are parallel has an answer. But if one is confined to look at the manifold only, like our imaginary ant, then no answer is possible without further structure.

One approach is to make our ant carry the vector from Chicago to New York so that they can be compared in the same tangent space. But the poor ant may turn the vector around as it drags it along and the comparison between the transported vector and the one in New York becomes arbitrary. To avoid this, we must give the ant a rule which prevents it from turning the vector around or a rule to transport the vector parallel to itself. The rule is the additional structure needed to

Figure 7

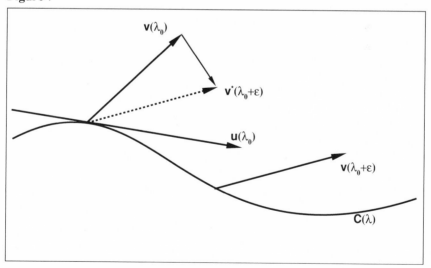

determine parallelism and is called the *affine connection*. A differential manifold may admit many connections but they cannot be totally arbitrary. A connection must satisfy certain linearity and derivative-of-product rule.

Consider a curve $C(\lambda)$ on a differential manifold with a connection and let $U = \partial / \partial \lambda$ [the tangent vector to $C(\lambda)$ at λ is $u(\lambda)$]. Let V be a vector field defined on the curve with $v(\lambda_0)$ the vector at λ_0 and $v(\lambda_0 + \varepsilon)$ the vector at $\lambda_0 + \varepsilon$. Let $v^*(\lambda_0 + \varepsilon)$ be $v(\lambda_0 + \varepsilon)$ parallel transported to λ_0. Then the covariant derivative can be defined as

$$\nabla_U V = \frac{v^*(\lambda_0 + \varepsilon) - v(\lambda_0)}{\varepsilon} \tag{15}$$

If (U, f) be a proper coordinate chart at P, then the covariant derivative can be expressed using Christoffel symbol Γ_{ij}^k, a scalar. First consider the covariant derivative of basis vectors.

$$\nabla_{\partial_j} \partial_i = \sum_k \Gamma_{ij}^k \frac{\partial}{\partial x^k} \tag{16}$$

where $U = \partial / \partial x^j = \partial_j$ and $V = \partial_i$

The expression can be extended to the covariant derivative of any tensor field since tensors are all expressible as linear combination of basis tensors and they (vectors and 1-form) are derivable from the vector basis.[10]

Given any connection, we can define a geodesic as the curve which parallel transports its tangents.[11] This is the closest thing, on the manifold, to a straight line which we intuitively associate with the shortest path between two points. But we have not used the metric! If a distance is defined on the manifold, it will define its own set of shortest curves, the metric geodesic. A connection is compatible with metric if the two geodesics coincide; the condition is that the covariant derivative of the metric tensor is identically zero.

2.8 Groups in Differential Equations

We already mentioned the abstract definition of Lie groups in footnote 6. One of Lie's main results is that it is possible to assign to every Lie group a much simpler algebraic object, the Lie algebra. The converse question, whether a Lie group can be associated with a Lie algebra is of great importance and is discussed in Spivak (1979). We will restrict ourselves to the application of Lie groups to differential equations and, to simplify matters, confine to the case of two variables and one parameter.[12]

A set of transformations

$$x' = \phi(x, y, t) \qquad y' = \psi(x, y, t) \tag{17}$$

is said to form a parameter group if the results of successive transformations is an element of the group.[13] In terms of the definition of a group given in section I, group property requires:

1. There exists a t'' depending on t and t' such that

$$\phi(x', y', t') = \phi(x, y, t'') \qquad \psi(x', y', t') = \psi(x, y, t'')$$

2. There exists a t_0 such that

$$x = \phi(x, y, t_0) \qquad y = \psi(x, y, t_0)$$

3. There exists a t_{-1} such that

$$x = \phi(x', y', t_{-1}) \qquad y = \psi(x', y', t_{-1})$$

We assume that ϕ and ψ are real valued analytic functions.[14] We have already mentioned in section I that Lie's theory permits us to determine the invariance of an integral of a differential equation without knowing the integral. This is a consequence of the fact that continuous groups of functions, ϕ and ψ are completely determined by the values of its partial derivatives at t_0. Without loss of generality, we can take $t_0 = 0$.

Define an *infinitesimal operator*

$$U = \left(\frac{\partial \phi}{\partial t}\right)_0 \frac{\partial}{\partial x} + \left(\frac{\partial \psi}{\partial t}\right)_0 \frac{\partial}{\partial y} = \xi \frac{\partial}{\partial x} + \eta \frac{\partial}{\partial y} \qquad (18)$$

where $\xi = \left(\dfrac{\partial \phi}{\partial t}\right)_0$ and $\eta = \left(\dfrac{\partial \psi}{\partial t}\right)_0$. Note also that $\xi = Ux$ and $\eta = Uy$.

The effect of a finite transformation in parameter from $t_0 = 0$ to t can be derived as follows. Expanding by Maclaurin's series,

$$x' = \phi(x, y, t) = x + t\left(\frac{\partial \phi}{\partial t}\right)_0 + \frac{t^2}{2!}\left(\frac{\partial^2 \phi}{\partial t^2}\right)_0 + \dots$$

Now $\phi(x, y, 0) = x$, $\psi(x, y, 0) = y$ and $(\partial \phi / \partial t)_0 = Ux$ so that

$$\left(\frac{\partial^2 \phi}{\partial t^2}\right)_0 = \left[\frac{\partial}{\partial t}\frac{\partial \phi}{\partial t}\right]_0 = \left[\frac{\partial}{\partial x}\left(\frac{\partial \phi}{\partial t}\right)\frac{\partial \phi}{\partial t} + \frac{\partial}{\partial y}\left(\frac{\partial \phi}{\partial t}\right)\frac{\partial \psi}{\partial t}\right]_0$$

$$= \left[U\left(\frac{\partial \phi}{\partial t}\right)\right]_0 = UUx = U^2x$$

and so on. Similarly

$$\psi(x, y, t) = y + tUy + \frac{t^2}{2!}U^2y + \dots$$

This shows that the value of function $f(x, y)$ at (x', y') is determined by Ux and Uy.

So far we derived the infinitesimal transformation knowing the group functions ϕ and ψ. The converse problem is to determine these functions from the infinitesimal generators. Given the infinitesimal transformation

$$\delta x = Ux \cdot \delta t = \xi \cdot \delta t \qquad \delta y = \eta \cdot \delta t, \tag{19}$$

the point (x, y) is carried to a neighboring point $(x + \xi\delta t, y + \eta\delta t)$. The repetition of this process a number of times is equivalent to moving along the integral curve of the system of differential equations[15]

$$\frac{dx'}{\xi(x', y')} = \frac{dy'}{\eta(x', y')} = \frac{dt}{1} \tag{20}$$

The first two give an integral of the form

$$u(x', y') = \text{constant}$$

and from the second pair, we can find a solution

$$v(x', y') - t = \text{constant}$$

From these simultaneous equations, the value of x' and y' can be determined and shown to satisfy the group properties.

The infinitesimal transformations can be used to define invariances of a function, of a set of curves and of an equation, to the group of transformations. A function $f(x, y)$ is *invariant* under the group of transformations of x and y if $f(x, y) = f(x', y')$. Expanding by Taylor's series and using infinitesimal transformations, this can be written as

$$f(x', y') = f(x, y) = f(x, y) - t \cdot Uf + \frac{t^2}{2!} U^2 f + \dots$$

so that the necessary and sufficient condition for invariance is that $Uf \equiv 0$. Notice that the condition involves only the infinitesimal transformations.

A family of curves $C(x, y)$ = constant, is invariant if $C(x', y')$ = constant generates the same set of curves. This is satisfied if for each value of t, a curve of $C(x, y)$ = constant, is mapped by the transformation (17) to another curve. If the family of curves, $C(x, y)$ = constant, is considered to be the level sets of a function (e.g., indifference curves are level sets of a utility function), then the transformation leaves the level sets invariant, making it a symmetry of the system.

The condition for the invariance of a set of curves is that the transformation maps each curve to another curve in the same set

$$C(x', y') = h\{C(x', y')\}$$

Expanding by Taylor series,

$$C(x', y') = C(x, y) + t \cdot UC + \frac{t^2}{2!} U^2C + \dots$$

The necessary and sufficient condition for the invariance of the family of curves is that UC should be a function of C alone (i.e., not of the coordinates of the individual points).

$$U(C) = G(C) \tag{21}$$

Finally an equation $F(x, y) = 0$ is invariant if $F(x', y') = 0$ whenever $F(x, y) = 0$. A necessary and sufficient condition is that $UF = 0$.

A transformation (17) carries with it the transformation of $p = dy / dx$, as

$$\frac{dy'}{dx'} = p' = \frac{\dfrac{\partial \psi}{\partial x} dx + \dfrac{\partial \psi}{\partial y} dy}{\dfrac{\partial \phi}{\partial x} dx + \dfrac{\partial \phi}{\partial y} dy} = \frac{\dfrac{\partial \psi}{\partial x} + \dfrac{\partial \psi}{\partial y} p}{\dfrac{\partial \phi}{\partial x} + \dfrac{\partial \phi}{\partial y} p}.$$

We now have the once extended transformation

$$x' = \phi(x, y, t), \quad y' = \psi(x, y, t), \quad p' = \chi(x, y, p, t) \tag{22}$$

The differential equation

$$L(x, y, p) = 0 \tag{23}$$

is invariant under the transformation (17) (i.e., one integral curve is mapped to another) if and only if it is invariant under (22). The condition for this is (compare it with the condition for the algebraic equation $F(x, y) = 0$ above)

$$U'L \equiv \xi \frac{\partial L}{\partial x} + \eta \frac{\partial L}{\partial y} + \eta' \frac{\partial L}{\partial p} = 0$$

where

$$\eta' = \frac{\partial \eta}{\partial x} + \left(\frac{\partial \eta}{\partial y} - \frac{\partial \xi}{\partial x} \right) p - \frac{\partial \xi}{\partial y} p^2$$

Consider a curve $F(x, y) = 0$. If (x, y) is a point on the curve, p is the slope of the tangent to the curve. The transformation (17) takes (x, y) to (x', y'), $F(x, y)$ to $F(x', y')$ and p to p'. But p' depends only on x, y, and p; so any curve tangent to $F(x, y) = 0$ at (x, y) will be transformed into a curve tangent to $F'(x', y') = 0$ at (x', y'). The extended transformation tells us not only how indifference curves are transformed but how their slopes are transformed by (17).[16]

2.9 Calculus of Variation and The Hamiltonian Formulation

Earlier we defined a curve as a mapping from an open interval of a real line to a manifold. In this section we consider the choice of curves that maximize an integral. First we derive the Euler equations for the simplest case where the system has one degree of freedom[17]; taking the parameter to be t, we want to choose a curve $x(t)$, $t = a$ to $t = b$, that maximize the integral $\int F (t, x, \dot{x}) dt$ where $\dot{x} = dx / dt$. A well-known example is the choice of consumption expenditure that maximizes inter-temporal utility. Then we will state its extension to systems with n degrees of freedom and, reformulate the problem using the Hamiltonian.

Divide the interval, $t = a = t^0$ to $t = b = t^{n+1}$, into $(n+1)$ equal parts of length $\Delta t, (t^0, t^1), \ldots (t^n, t^{n+1})$. The purpose is to approximate the integral by a sum; the process is similar to that used in elementary calculus to approximate the area under a curve. Additional complications arise from the fact that $F(\ )$ is not a function of t alone but also of x and $\dot{x}$. We first approximate $x(t)$ by a polynomial line with vertices $(t^0, x^0), (t^1, x^1), \ldots (t^{n+1}, x^{n+1})$. The slope of the line with vertices (t^i, x^i) and (t^{i+1}, x^{i+1}) is $(x^{i+1} - x^i) / \Delta t$. We use this expression to approximate $\dot{x} (t^i)$. Substituting we can approximate the integral by the sum

$$J' (t^0, t^{n+1}) = \sum_{i=0}^{n+1} F (t^i, x^i, \frac{x^{i+1} - x^i}{\Delta t}) \Delta t$$

The curve maximizes the sum if marginally changing x at t^i (see figure 8) leads to no variation in the sum. We examine the variation of the sum by partially differentiating it with respect to x^i.

Figure 8

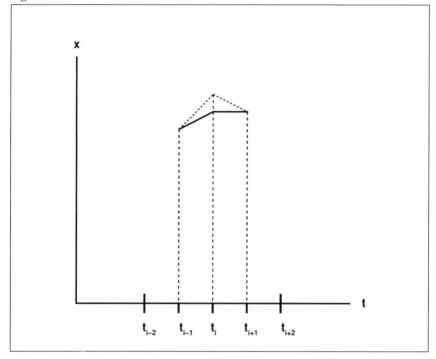

$$\frac{\partial J'}{\partial x_i} = F_x(\ )\ \Delta t + F_{\dot{x}}\left(t^{i-1}, x^{i-1}, \frac{x^i - x^{i-1}}{\Delta t}\right)$$

$$-F_{\dot{x}}\left(t^i, x^i, \frac{x^{i+1} - x^i}{\Delta t}\right)$$

Dividing both sides by Δt,

$$\frac{\partial J'}{\partial x_i\, \Delta t} = F_x(\ ) + \frac{F_{\dot{x}}(t^{i-1}, \ldots) - F_{\dot{x}}(t^i, \ldots)}{\Delta t}$$

As $\Delta t \to 0$, this expression converges to the limit called the variational derivative of $J\,(t)$, given by

$$\frac{\delta J}{\delta x} = F_x\,(t, x, \,\overset{\circ}{x}\,) - \frac{\partial}{\partial t}\,F_{\overset{\circ}{x}}\,(t, x, \,\overset{\circ}{x}\,)$$

The integral reaches an extremal if this variational derivative is zero and thus we derive the second order differential equation known as Euler equation

$$F_x - \frac{d}{dt}\,F_{\overset{\circ}{x}} = 0$$

If the system has n degrees of freedom so that $x(t) = (x^1(t), ..., x^n(t))$, then we have n Euler equations

$$F_{x_i} - \frac{d}{dt}\,F_{\overset{\circ}{x}_i} = 0 \tag{24}$$

The n second order partial differential equations can be reduced to a system of $2n$ first order partial differential equations by defining a set of $2n + 1$ canonical variables $t, x^1, ..., x^n, p^1, ..., p^n$ related to the earlier set of variables by the equations

$$t = t, \quad x^i = x^i, \quad p^i = F_{\overset{\circ}{x}^i}$$

In economic models, p^i is interpreted as shadow prices. Now write a new function called the Hamiltonian

$$H = -F + \sum_i \overset{\circ}{x}^i p^i \tag{25}$$

We shall now seek to express the Euler equations in terms of "canonical variables." Computationally the most convenient approach (Gelfand and Fomin, 1963, p. 69) is to take the partial derivatives of the function H

$$dH = -dF + \Sigma p^i \, d\dot{x}^i + \Sigma \dot{x}^i dp^i$$

$$= -\frac{\partial F}{\partial t} \, dt - \Sigma \frac{\partial F}{\partial x^i} \, dx^i - \Sigma \frac{\partial F}{\partial \dot{x}^i} \, d\dot{x}^i + \Sigma p^i \, d\dot{x}^i + \Sigma \dot{x}^i \, dp^i$$

$$= -\frac{\partial F}{\partial t} \, dt - \Sigma \frac{\partial F}{\partial x^i} \, dx^i + \Sigma \dot{x}^i dp^i$$

Partial differentials of H can now be shown to be given by the coefficients of the differentials on the right-hand side

$$\frac{\partial H}{\partial t} = -\frac{\partial F}{\partial t}, \qquad \frac{\partial H}{\partial x^i} = -\frac{\partial F}{\partial x^i}, \qquad \frac{\partial H}{\partial p^i} = \dot{x}^i \qquad (26)$$

The Euler equation can now be written as

$$\frac{dx^i}{dt} = \frac{\partial H}{\partial p^i}, \qquad \frac{dp^i}{dt} = -\frac{\partial H}{\partial x^i}, \quad i = 1, \ldots, n.$$

2.10. Conservation Laws and Noether Theorems

In dynamic models, we want to study the motion of the system over time. One approach, "the vectorial method," seeks to study the forces acting on the system when it is at a point and determine the direction of motion there. The integration of the resulting differential equation, subject to the appropriate boundary value conditions, gives the locus of motion. In an alternate approach, "analytical mechanics," we seek to characterize the entire trajectory as that which minimizes an action integral.

The idea that the dynamics of a natural phenomenon can be explained by the minimization of some action integral and that it reflects the simplicity inherent in the system has a long tradition. Yourgrau and Mandelstom (1979, p. 4) traces it back to a passage in Aristotle. Hero of Alexandria established a genuinely scientific minimum principle when he showed that the path of light reflected by a mirror takes the shortest possible path to the observer's eyes. The law is a direct antecedent of Fermat's principle of least time. Maupertius proclaimed it to be a universal law in the spirit of Platonic-Pythagorian cosmology and so made it a controversial proposition. It was left to the English physicist, Sir William R. Hamilton, to put analytical dynamics on a solid mathematical foundation.

We shall now show how an infinitesimal transformation of a (n+1) space, ($t, x^1, ..., x^n$) affects the action integral of a dynamic system and how the assumptions of the invariance of an integral leads to conservation laws. Logan (1977), Lovelock and Rund (1975) and Sato (1981) give detailed and rigorous mathematical analysis of the invariance identities. Here we will avoid purely technical details; after stating the Noether theorem, we shall provide an intuitive explanation of the special case when t does not appear explicitly in the Hamiltonian H in equation (25).

Unlike in section 2.7, we will consider transformations that depend on r essential parameters, $\varepsilon = (\varepsilon^1, ..., \varepsilon^r)$. The number of conservation laws depend on the number of parameters and arbitrarily setting it equal to one affects the analysis substantially. Let the transformation be given by the equations

$$\bar{t} = \phi(t, x, \varepsilon)$$

$$\bar{x}^k = \psi^k(t, x, \varepsilon), \quad k = 1, ..., n, \quad x = (x^1, ..., x^n)$$

(27)

The identity transformations are given by $\varepsilon = 0$. Expanding the right-hand side of (26) by Taylor series about $\varepsilon = 0$, we get

$$\overline{t} = t + \sum_s \tau_s (t, x) \, \varepsilon^s + O(\varepsilon) \qquad s = 1, \dots, r.$$

$$\overline{x}^k = x^k + \sum_s \xi_s^k (t, x) \, \varepsilon^s + O(\varepsilon)$$

The principal linear parts, τ_s and ξ_s^k, are the infinitesimal generators of the transformation (26) and is given by

$$t_s (t, x) = \frac{\partial \phi}{\partial \varepsilon^s} (t, x, 0) \quad \text{and} \quad \xi_s^k (t, x) = \frac{\partial \psi^k}{\partial \varepsilon^s}. \tag{28}$$

Consider any curve $x(t)$, $t = a$ to $t = b$, in the original configuration space $(t, x^1, \dots, x^n)$. The transformation of the space to $(\overline{t}, \overline{x}^1, \dots, \overline{x}^n)$ maps $x(t)$ to $\overline{x}(\overline{t})$. The integrand referred to as Lagrangian, $L(t, x(t), \dot{x}(t))$ is similarly transformed to $L\{\overline{t}, \overline{x}(\overline{t}), (d\overline{x}(\overline{t})/d\overline{t})\}$. Finally, the action integral

$$J(x) = \int_a^b L(t, x(t), \dot{x}(t)) \, dt \tag{29}$$

is transformed to

$$J(x) = \int_{\overline{a}}^{\overline{b}} L(\overline{t}, \overline{x}(\overline{t}), \frac{d\overline{x}(\overline{t})}{d\overline{t}}) \, d\overline{t}$$

We will now state the Noether theorem, skipping the derivation. Define

$$E_k = \frac{\partial L}{\partial x^k} - \frac{d}{dt} \frac{\partial L}{\partial \dot{x}^k}, \quad k = 1, \dots, n. \tag{30}$$

E_k is the left hand side of the Euler equation (24); Noether identities state that if the integral is invariant to the transformation, then there are linear combinations of E_k that can be expressed as exact differentials, $d\Omega_s / dt$ of a function Ω_s, where

$$\Omega_s = \left(L - \sum_k \overset{\circ}{x}{}^k \frac{\partial L}{\partial x^k} \right) \tau_s + \frac{\partial L}{\partial x^k} \xi_s - \Phi_s,$$

$$= -H\tau_s + \sum_k \frac{\partial L}{\partial \overset{\circ}{x}{}^k} \xi_s^k - \Phi_s,$$

$$k = 1, \ldots, n; \ s = 1, \ldots, r. \quad (31)$$

If $\Phi_s \equiv 0$, then the fundamental integral (29) is *absolutely invariant*; if Φ_s has to be determined in the process of solving the system, then the integral is said to be *divergence invariant.*

But along an extremal, E_k is equal to zero and so $(d / dt)\Omega_s = 0$ implying that Ω_s is a constant. This leads to the Noether theorem.

Noether Theorem: If the action integral of a problem is invariant under the r-parameter family of transformation, then r distinct quantities Ω_s, s = 1, ..., r, are constant along any extremal.

In other words, Ω_s are the constants we were looking for or $\Omega_s = $ constant are the conservation laws. An intuitive understanding of the result can be obtained by considering the special case where the objective is to maximize the integral

$$J(x) = \int_a^b L(x^1, \ldots, x^n, \overset{\circ}{x}{}^1, \ldots, \overset{\circ}{x}{}^n) \, dt$$

where L does not depend on t explicitly. In this case, the corresponding Hamiltonian (equation (25)) is not an explicit function of t. It is clear that L and the integral remains invariant if we replace t with a new variable $t+\varepsilon$ for an arbitrary ε. In fact, given a curve $x(t)$, the transformed curve is given $x(t^* - \varepsilon) = x^*(t^*)$ with the appropriate limits of integration. So

$$J(x^*) = \int_{a+\varepsilon}^{b+\varepsilon} L(x^{1*}, ..., x^{n*}, \overset{\circ}{x}{}^{1*}, ..., \overset{\circ}{x}{}^{n*}) \, dt^*$$

$$= \int_{a+\varepsilon}^{b+\varepsilon} L(x^1(t^* - \varepsilon), ..., \overset{\circ}{x}{}^n(t^* - \varepsilon)) \, d(t^* - \varepsilon)$$

$$= \int_a^b L(x^1(t), ..., \overset{\circ}{x}{}^n(t)) \, dt$$

$$= J(x)$$

The transformation can be written as $t^* = t + \varepsilon$, $x^{s*} = x^s$, $s = 1, ..., n$. The corresponding infinitesimal transformation is $t = 1$ and $\xi^s = 0$. In this case, equation (31) can be written as

$$\Omega_s = -H \cdot 1 + \frac{\partial L}{\partial x_s} \cdot 0 = -H = \text{a constant.} \qquad (32)$$

When t does not appear explicitly in $L(\ )$ or H, then the Hamiltonian itself is conserved. This special case of great importance in economics.

III. HOLOTHETICITY: Symmetry of the Isoquant Map

Isoquants or level sets of a production function establish a relation between the quantities of inputs and that of output. In section I, we pointed out that technical progress can be viewed either as a shift in production itself or as an increase in inputs measured in efficiency units. If technical progress is Hicks-neutral, then the input vector OP_1

in figure 9 acts as if it is OP_1'. The example leads to two questions: What way can this concept of technical progress be generalized? What are the common restrictions on the functions needed to make them meaningful?

If $\overline{K}$ and $\overline{L}$ are inputs in efficiency units, then it can be expressed in natural units as,

$$\overline{K} = A\ (t)\ K = \lambda_1\ (t)\ K, \quad \overline{L} = A\ (t)\ L = \lambda_2\ (t)\ L \qquad (33)$$

where $\lambda_1 = \lambda_2 = A(t)$. The standard case of factor augmenting biased technical progress is

$$\overline{K} = \lambda_1\ (t)\ K \qquad \overline{L} = \lambda_2\ (t)\ L$$

We can relax the assumption that λ_i depends only on t. The next generalization is to make it a function of K / L also; the rate of technical progress differs from a ray to another. A further generalization is to make it a function of K and L. We can classify these cases as

	Case 1		Case 2	
	$\lambda_1 = \lambda_2$		$\lambda_1 \neq \lambda_2$	
	$\lambda_1\ (t) = \lambda_2\ (t)$		$\lambda_1\ (t) \neq \lambda_2\ (t)$	(34)
	$\lambda_2\ (t, K/L) = \lambda_2\ (t, K/L)$		$\lambda_1\ (t, K/L) \neq \lambda_2\ (t, K/L)$	
	$\lambda_1\ (t, K, L) = \lambda_2\ (t, K, L)$		$\lambda_2\ (t, K, L) \neq \lambda_2\ (t, K, L)$	

The common restriction needed is that the impact of technical progress in period $t^2 - t^0$ should equal the sum of changes in period $t^1 - t^0$ and $t^2 - t^1$. If this restriction is not met, we cannot unambiguously speak of the efficiency at time t^2 relative to time t^0. Similarly we want the $\lambda_i(\ , 0) = 1$. Finally we want to say that if the inputs have higher efficiency at time t^1 relative to time t^0, the reverse is true at time t^0 relative to time t^1. These elementary considerations suggest that the scalar functions λ_i must satisfy group properties. Consider the case where

$$\lambda_1(t) = \lambda_2(t) = e^{\alpha t}$$

For this case, the infinitesimal operator is

$$U = \alpha K \frac{\partial}{\partial L} + \alpha L \frac{\partial}{\partial L} \qquad (35)$$

since $(dK / dt)_0 = \alpha K = UK$ and $(dL / dt)_0 = \alpha L = UL$. Expanding the technical progress function using Taylor series

$$K' = K + UK \cdot t + U^2 K \frac{t^2}{2!} + \dots \qquad (36)$$

$$L' = L + \alpha L \cdot t + U^2 L \frac{t^2}{2!} + \dots \qquad (37)$$

We can interpret this process economically as follow. A firm which employs an input vector (K_1, L_1) produces an output $F(K_1, L_1)$ initially. Over a period of time t, the factors increase in efficiency, so that they now have the productive capacity of the vector $(e^{\alpha t} K_1, e^{\alpha t} L_1)$. We can figuratively speak of a *virtual expansion path* in the input space as given by the infinite series (36).

Now consider the process starting from another input combination (K_2, L_2) on the same isoquant. Technical progress will transform it to $(e^{\alpha t} K_2, e^{\alpha t} L_2)$. If the production function is holothetic to the technical progress or is a symmetry of the isoquant map, then P'_1 and P'_2 must lie on the same isoquant.

Starting from a linear homogeneous production function, we can enquire whether there is a transformation of the production surface that maps one isoquant to the other the same way as the technical

Figure 9

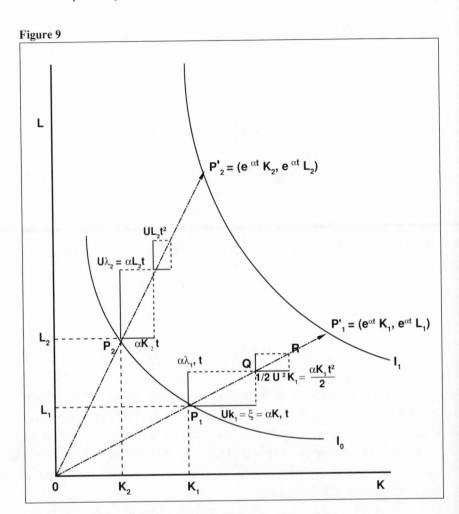

progress function did. If it does, the economies of scale corresponding to the transformed production function has the same effect on producti ity as the technical progress itself. The possible transformation can be obtained by solving (see equation (21))

$$Uf = \xi \frac{\partial f}{\partial K} + \eta \ \frac{\partial f}{\partial L} = G(f) \tag{38}$$

The solution of (38) can be derived from the corresponding system of ordinary differential equation

$$\frac{dK}{\xi} = \frac{dL}{\eta} = \frac{df}{G(f)} \qquad (39)$$

Substituting $\xi = \alpha K$ and $\eta = \alpha L$ for the Hicks-neutral case, it can be shown that the function holothetic to Hicks-neutral technical progress is the homothetic function (Sato, 1981, pp. 30–34). This was the conclusion derived in Sato and Ramachandran (1974) but the new method is not confined to Hicks neutral case.

In the case of biased technical progress with $\lambda_1 (t) = e^{\alpha t}$ and $\lambda_2 (t) = e^{\beta t}$, the infinitesimal generators are $\xi = \alpha K$ and $\eta = \beta L$. Substituting in (39), the holothetic production function is seen to be

$$Y = F(K^{1/\alpha} Q(L^{\alpha} / K^{\beta})) \qquad (40)$$

The almost homogeneous production function proposed by Lau (1978) and the almost homothetic function of Sato (1977) are special cases of (40). Sato (1981) examines many other technical progress functions and their corresponding holothetic production functions.

IV. EXAMPLES OF CONSERVATION LAWS IN ECONOMICS

Consider a Solow-Swan type economy with an aggregate production function

$$Y = F(K, L) \qquad (41)$$

Labor grows at an exogenous rate $\dot{L} / L = n$. Then per capita output, per capita capital and per capita consumption are given by

$$y = \frac{Y}{L}, \quad k = \frac{K}{L} \text{ and } c = \frac{C}{L}$$

Further

$$\frac{\dot{k}}{k} = \frac{\dot{K}}{K} - \frac{\dot{L}}{L} = \frac{Y - C}{K} - n = \frac{Y}{K} - \frac{C}{K} - n$$

Hence $\dot{k} = y - c - nk$ or $c = c(k, \dot{k}) = y - nk - \dot{k}$

The objective of the society is to maximize utility from consumption over time. In a classical paper, Ramsey (1928) considered a special case of this model with $n = 0$ and L normalized to 1. Ramsey suggested that the appropriate criteria for the economy is to maximize

$$J(C) = \int_0^\infty (B - U(C))\, dt$$

$$\text{subject to } c = C/1 = F(K) - \dot{K} \quad (42)$$

where B is the upper bound of $U(\ )$ is called the bliss point. The problem can be reformulated using the Hamiltonian (Note that a minimization problem can be converted into maximization problem by multiplying by -1)

$$H = -(B - U(C) + p(t)\dot{K}$$

$$= -(B - U(C)) + p(t)(F(K) - C) \quad (43)$$

The control variable, C, is selected so as to maximize H [see Burmeister and Dobbell (1970) for a discussion of the details of control theory and the Ramsey problem].

$$\frac{\partial H}{\partial C} = U'(C) - p(t) = 0$$

$$U'(C) = p(t)$$

Substituting in the Hamiltonian which is not explicitly a function of t

$$-(B - U(C)) + U'(C) K = \text{constant by (32)}$$

$$\dot{K} = \frac{B - U(C) + \text{constant}}{U'(C)} \tag{44}$$

An economic interpretation of this rule can be obtained by considering the loss in present utility due to a contemporaneous increase in capital formation with the added consumption flow that the resulting capital formation would generate. This approach was proposed by Keynes. If capital formation is increased at time t by h for a period Δt, then the loss in utility due to reduced consumption is approximately $hU'(C) \Delta t$. The increased output at any point of time is $hF'(K) \Delta t$ and the present value of this flow is

$$\int_0^\infty h \, \Delta t \, F'(K) \, U'(C) \, dt$$

Hence, equating the extra cost to extra benefit,

$$\int_0^\infty h \, \Delta t \, F'(K) \, U'(C) \, dt = hU'(C) \, dt$$

Differentiating with respect to t

$$- F'(K) U'(C) = \dot{U}'(C)$$

This is nothing but the Euler equation of a variational problem stated earlier, Max $J(C)$ with $C = F(K) - \dot{K}$. Since the integrand does not contain t explicitly, it was shown that the first integral of the Euler equation is

$$B - U = KU'(C) + \text{constant}$$

which takes us back to the Ramsey rule.

While Ramsey's model has one sector and an objective function that seeks to maximize the intertemporal utility from consumption, von Neumann model has n sectors but no consumption.[18] Hence the rate of increase of the stock of any good equals its output. The vector of capital goods at time t is $K_t = (K_t^i)$ and is used to produce the vector of net capital formation which equals output. The transformation function relating K_t to $\dot{K}_t$ is

$$F(K_t, \dot{K}_t) = 0$$

We assume that F is homogeneous of degree one, concave and smoothly differentiable. The process is considered to have begun at time $t = 0$ with capital vector K_0 and terminates at time T. If the terminal vector $K_t' = (K_t^2, ..., K_t^n)$ is given, then the objective is to maximize K_t^1 subject to K_0 and K_t.

The variational problem in continuous time can be written as

$$\text{Max} \int_0^\infty K_t^1 \, dt \quad \text{subject to} \quad F(K_t, \dot{K}_t) = 0$$

$$\text{and the boundary conditions.} \quad (45)$$

Since the integral does not contain t, explicitly, the Hamiltonian is constant.

$$H = L_t - \sum_j \dot{K}^j_t \frac{\partial L_t}{\partial \dot{K}^j_t}$$

$$= K^1_t + \lambda_t F - \sum_j \dot{K}^j_t \frac{\partial(\dot{K}_t - \lambda_t F)}{\partial \dot{K}^j_t}$$

$$= -\lambda_t \sum \dot{K}^j_t \frac{\partial F}{\partial \dot{K}^j_t}$$

Samuelson (1970), in deriving the first explicit statement of a conservation law, shows that $H = -\lambda_t Y_t$. It follows that

$$\lambda_t Y_t = \text{constant} \qquad (46)$$

Samuelson also derived another conservation law involving wealth but no general principle like the constancy of the Hamiltonian for the first law, was evident. Sato (1981) showed that a second law can be derived, after lengthy manipulations from the principle of divergence-invariance

$$\lambda_t W_t = \text{constant} \qquad (47)$$

Taking the ratios of the two conservation laws to eliminate λ_t, the unobservable shadow price, we obtain a constancy of capital-output ratio

$$\frac{W_t}{Y_t} = \text{constant} \qquad (48)$$

Chapters 2 to 6 deal derive conservation laws for different economic models.

The insight of symmetry analysis is in the recognition that conservation laws are not *ad hoc* derivations of the constancy of a scalar function. It is intimately connected with the structure of the model. The Ramsey conservation law and the first conservation law arise from a transformation of the time axis through a shift in the origin. In other words, time in economic models has no distinguished origin. The second conservation law has infinitesimal generators that are constant multiples of the inputs; it is a transformation of the units of inputs.

V. CONCLUSION

The differential geometric approach provides a unified methodological framework for invariance, symmetry and conservation laws. Its strength lies in examining the informational implications of the assumptions behind economic models. Do we have any information about production functions over and above that is in the isoquant map? If not, how are we to distinguish between transformations that map the isoquants isomorphically? Conservation laws, on the other hand, arise from the transformations which leave the dynamics of a model invariant.

This volume seeks to explore varied applications of these techniques.

Notes

1. Misner, Thorne and Wheeler (1973, p. 48). For a description of the Erlanger Program, see Yaglom (1988) and Klein (1972, pp. 917–921).
2. Like all grand visions, the Erlanger Program had its limitations. See the references in note 1, above.
3. See Friedman (1983) for a recent and somewhat critical review of the inter-action between relativity physics and positivist philosophy.
4. Lie algebra and the topological implications of Lie's work were actively re-searched during this period.
5. An n-dimensional topological manifold M^n is a Hausdorff topological space with a countable basis for the topology, which is locally homeomorphic to R^n. A differentiable manifold is a topological manifold with a differential structure. See Bröker and Jänich (1973, pp. 1–4).
6. A Lie group is a group X which is also a manifold with C^∞ structure such that $(x, x') \to xx'$ and $x \to x^{-1}$ are C^∞ functions (Spivak, 1979, p. 501). Just as there are function spaces, there are manifolds of transformation. They should be differentiated from manifolds of the underlying space. The reader is referred to Dubrovin, Fomenko, and Novikov (1984, 1985) and Spivak (1979) for a formal definition of the other concepts discussed in this section.
7. A reader who is not familiar with the approach would enjoy the aside in Misner, Thorne and Wheeler (1973, p. 227):

   "Tangent vectors equal directional derivative operator? Preposterous! A vector started out as a happy, irresponsible trip from $\mathbb{P}_0$ to $\mathbb{Q}_0$. It ended up with the social responsibility to tell how something changes at $\mathbb{P}_0$. At what point did the vector get saddled with the unexpected load? And did it really change its character all that much, as it seems to have done?"

   The answer to the last question is no; otherwise this whole exercise is meaningless.

   A technical note: many curves may be tangent to each other at P and the tangent vector is defined in terms of the equivalence classes of curves.
8. A clear discussion of these niceties, using a special manifold E^n, is in Edelen (1985, pp. 26–27).
9. Misner, Thorne and Wheeler (1973, pp. 74–76) and Schutz (1980, p. 57).
10. An alternate definition of the Christoffel symbol is based on differentiation of an arbitrary tensor. Consider for simplicity, the vector field T^i on a space of coordinates $(x^1, ..., x^n)$. Then the quantities $T^i_s = \partial T^i / \partial x^s$ transforms like the components of a $\binom{1}{1}$ tensor for all linear coordinate changes $x^i = \sum_j a^i_j z^j$. Hence

$$\bar{T}^{j}_{;1} = \frac{\partial T^{j}}{\partial z^{1}} = \sum_{i,p} T^{i}_{;p} \frac{\partial x^{p}}{\partial z^{1}} \frac{\partial z^{j}}{\partial x^{i}}$$

where $\bar{T}^{j}$ is the component in $(z^{1}, ..., z^{n})$. But if the coordinate transformation is nonlinear, then

$$T^{j}_{;1} = \frac{\partial T^{j}}{\partial z^{1}} + \sum_{s} \Gamma^{j}_{s1} T^{s}$$

where

$$\Gamma^{j}_{s1} = -\sum_{i,m} \frac{\partial x^{i}}{\partial z^{s}} \frac{\partial x^{m}}{\partial z^{1}} \frac{\partial^{2} z^{j}}{\partial x^{i} \partial x^{m}} \quad , \text{the Christoffel symbol;}$$

note that Γ^{j}_{s1} vanishes when the transformation is linear. For a detailed treatment of this approach, see Dubrovin, Fomenko and Novikov (1984, pp. 271–295).

11. Another approach is to begin with the geodesic ("line of free fall") and define parallel transport and covariant derivative. See Misner, Thorne and Wheeler (1973, pp. 248–249).

12. In this section, we basically follow the excellent discussion of one parameter group by Cohen (1931). For a compact summary of r-parameter group, see Sato (1981).

13. Transformations form a manifold and the two definitions of Lie groups are consistent. See Dubrovin, Fomenko and Novikov (1984, pp. 120–136; 1985, pp. 10–15).

14. We assume that

$$\begin{vmatrix} \dfrac{\partial \phi}{\partial x} & \dfrac{\partial \phi}{\partial y} \\ \dfrac{\partial \psi}{\partial x} & \dfrac{\partial \psi}{\partial y} \end{vmatrix} \neq 0$$

A technical requirement is that the parameters must satisfy group composition properties; here t is a real number and the condition is satisfied.

15. For a simple but modern treatment of the system of equations, see Zachmanoglou and Thoe (1986). A diagrammatic treatment of the case of technical progress is given in section III below.

16. The extended transformation is a special case of contact transformations.

17. Gelfand and Fomin (1963) provides an excellent introduction to the Euler equations and the Hamiltonian formulation.

18. See Chapter 2 of this volume for Samuelson's classic paper on conservation law for the von Neumann model.

Reader's Guide

A reader who randomly picks a book on differential geometry or Lie groups in a library is in for a surprise. Mathematical textbooks adopt either the abstract or the computational approach. The former expects considerable "mathematical maturity" while the latter will look like a riot in indices. Application-oriented books will be full of examples from physical sciences. The following annotated list is intended to assist one with a background in economics or finance to get started; it is a subjective choice and not a ranking of available books.

1. Dubrovin, B. A., Fomenko, A. T. and Novikov, S. P. (1984, 1985). *Modern geometry—methods and applications.* New York: Springer Verlag.
    Part 1: The geometry of surfaces, transformation groups and fields.
    Part 2: The geometry and topology of manifolds.
    Another addition to the lucid textbooks that Russian mathematicians specialize in writing. The first volume has a chapter on calculus of variations. Because of leisurely presentation, manifolds appear only in the second volume.

2. Spivak, M. (1979). *A comprehensive introduction to differential geometry.* Wilmington, Delaware: Publish or Perish. (Volume 1, 2nd ed.)
    The great American differential geometry book provides an intelligible and authoritative account. The first chapter is devoted to manifolds and Chapter 10 to Lie groups.

3. Schutz, B. (1980). *Geometric methods in mathematical physics.* Cambridge: Cambridge University Press.
    Do not be misled by the title. Four of six chapters require no knowledge of physics and contain short and clear explanations of concepts that should be intelligible to one who followed this chapter.

4. Edelen, D.G.B. (1985). *Applied exterior calculus.* New York: John Wiley & Sons.
    A detailed discussion of differential forms. Part 1 discusses differential forms, Part 2 its applications in mathematics including calculus of variations, and Part 3 applications to physics.

5. Cohen, A. (1931). An *introduction to the Lie theory of one parameter groups.* New York: G. E. Stechert.
    An elementary book on Lie groups of differential equations that has no match.

6. Yaglom, I. M. (1988). *Felix Klein and Sophus Lie: Evolution of the idea of symmetry in the nineteenth century.* Boston: Birkhäuser.
    The subtitle is more descriptive of the book. An information-packed but occasionally rambling account of the mathematical tradition from Galois to Klein.

7.   Chalmers, A. F. (1982). *What is this thing called science?* St. Lucia: University of Queensland Press (2d ed.).
     A short, readable introduction to the many issues in the philosophy of science.

References

Alchain, A. (1953). The meaning of utility measurement. *American Economic Review, 42,* 26–50.

Bröcker, T. and Jänich, K. (1982). *Introduction to differential topology.* Cambridge: Cambridge University Press.

Burmeister, E. and Dobell, A. R. (1970). *Mathematical theories of economic growth.* London: The Macmillan Company.

Chalmers, A. F. (1982). *What is this thing called science?* (2d ed.) St. Lucia: University of Queensland Press.

Clapham, J. H. (1922). Of empty economic boxes. *Economic Journal, 32,* 305–314.

Cohen, A. (1931). *Introduction to the Lie theory of one parameter groups.* New York: G. E. Stechert & Co.

Dubrovin, B. A., Fomenko, A. T., and Novikov, S. P. (1984, 1985). *Modern geometry—methods and applications.* (Part 1: The geometry of surfaces, transformation groups, and fields. Part 2: The geometry and topology of manifolds). New York: Springer-Verlag.

Edelen, D.G.B. (1985). *Applied differential calculus.* New York: John Wiley & Sons.

Friedman, M. (1983). *Foundations of space-time theories.* Princeton, New Jersey: Princeton University Press.

Gelfand, I. M. and Fomin, S. V. (1963). *Calculus of variations.* Englewood Cliffs, New Jersey: Prentice-Hall.

Hicks, J. H. (1946). *Value and capital.* (2d ed.) London: Oxford University Press.

Klein, M. (1972). *Mathematical thoughts from ancient to modern times.* New York: Oxford University Press.

Lau, L. J. (1978). Application of profit functions. In M. Fuss and D. McFadden (eds.), *Production economics: A dual approach to theory and applications* (Volume 1). Amsterdam: North-Holland.

Logan, J. D. (1977). *Invariant variational principles.* New York: Academic Press.

Lovelock, D. and Rund, H. (1975). *Tensors, differential forms and variational principles.* New York: John Wiley & Sons.

Millman, R. S. and Parker, G. D. (1977). *Elements of differential geometry.* Englewood Cliffs, New Jersey: Prentice-Hall.

Misner, C., Thorne, K. S. and Wheeler, J. A. (1973). *Gravitation.* San Francisco: W. H. Freeman and Company.

Pigou, A. C. (1922). Empty economic boxes: A reply. *Economic Journal, 32,* 458–465.

Ramsey, F. P. (1928). A mathematical theory of saving. *Economic Journal, 38,* 543–559.

Reichenbach, H. (1960). *The theory of relativity and a priori knowledge.* Berkeley, California: University of Berkeley Press.

Sato, R. (1975). The impact of technical change on the holotheticity of production functions. Working paper presented at the World Congress of Econometric Society, Toronto.

Sato, R. (1977). Homothetic and nonhomothetic functions. *American Economic Review, 67,* 559–569.

Sato, R. (1981). *Theory of technical change and economic invariance.* New York: Academic Press.

Sato, R. and Ramachandran, R. (1974). Models of endogenous technical progress, scale effect and duality of production function. Providence, Rhode Island: Brown University, Department of Economics discussion paper.

Schutz, B. (1980). *Geometric methods of mathematical physics.* Cambridge: Cambridge University Press.

Solow, R. M. (1957). Technical change and aggregate production function. *Review of Economics and Statistics, 39,* 312–320.

Solow, R. M. (1961). Comment. In *Output, input, and productivity measurement* (Volume 25, Studies in income and wealth). Princeton, New Jersey: Princeton University Press.

Spivak, M. (1979). *A comprehensive introduction to differential geometry* (Volume 1). Wilmington, Delaware: Publish or Perish, Inc.

Stigler, G. J. (1961). Economic problems in measuring changes in productivity. In *Output, input, and productivity measurement* (Volume 25, Studies in income and wealth). Princeton, New Jersey: Princeton University Press.

Yaglom, I. M. (1988). *Felix Klein and Sophus Lie: Evolution of the idea of symmetry in the nineteenth century.* Boston: Birkhäuser.

Yourgrau, W. and Mandelstram, S. (1979). *Variational principles in dynamics and quantum theory.* New York: Dover.

Zellner, A. and Revankar, N. S. (1969). Generalized production functions. *Review of Economic Studies, 36,* 241–250.

Zachmangloe, E. C. and Thoe, D. W. (1976). *Introduction to partial differential equations with applications.* New York: Dover.

CHAPTER 2

Law of Conservation of the Capital-Output Ratio in Closed von Neumann Systems*

Paul A. Samuelson

Assumptions. A transformation function relates the vector of n capital goods, $K_t = (K_t^j)$, and the contemporaneous vector of net capital formations, $\dot{K}_t = (dK_t^j / dt)$, namely

$$F(K_t, \dot{K}_t) = 0 = C_t \qquad (1)$$

where F is assumed to be homogeneous-first-degree, concave, and smoothly differentiable. An optimal control program of intertemporal efficiency requires that, for assigned initial $(K_0^1, K_0^2, ..., K_0^n)$ and terminal $(K_T^2, ...K_T^n)$ at time $T > 0$, we must maximize K_T^1. This defines the variational problem

$$\text{Max} \int_0^T \dot{K}_t^1 \, dt \quad \text{subject to} \quad F(K_t, \dot{K}_t) = 0$$

and assigned boundary conditions. (2)

The Euler-Lagrange necessary conditions for this constrained-maximum problem are known to be given by

$$0 = \delta \int (\dot{K}_t^1 + \lambda_t F) \, dt = \delta \int L (K_t, \dot{K}_t, \lambda_t) \, dt \qquad (3)$$

*Aid from the National Science Foundation is gratefully acknowledged.

where the integrand L does not involve t explicitly. In consequence, any motion must satisfy the "energy integral"

$$L - \sum_{j=1}^{n} K^j \frac{\partial L}{\partial \dot{K}^j} \underset{t}{\equiv} \text{constant} = a$$

(4)

$$\dot{K}_t^1 - 0 - \dot{K}_t^1 + \lambda_t \sum_{1}^{n} \dot{K}_t^j \frac{\partial F}{\partial \dot{K}_t^j} \underset{t}{\equiv} a$$

Economic Interpretation. We can now interpretate this conservation law with C as *numeraire*. We define in the usual fashion[2]

$-\partial F / \partial \dot{K}^i = P^i$, the cost-price of the ith capital good

$\partial F / \partial K^i = R^i$, the net rental of the ith capital good

$Y_t = \sum_{1}^{n} P^j \dot{K}^j = \sum_{1}^{n} R^j K^j$, national income

$W_t = \sum_{1}^{n} P^j K^j$, national wealth   (5)

$r_t = \dot{W}_t / W_t$, the interest rate when $C_t \equiv 0$

$= (\Sigma P^j \dot{K}^j / \Sigma P^j K^j) + (\Sigma \dot{P}^j K^j / \Sigma P^j K^j)$

$= [\Sigma R^j K^j / \Sigma P^j K^j] + [\Sigma (\dot{P}^j / P^j) P^j K^j] / \Sigma P^j K^j$

$= \Sigma [(R^j / P^j) + (\dot{P}^j / P^j)] P^j K^j / \Sigma P^j K^j$

But the Euler variational conditions for a maximum can be shown to imply that *every* bracketed factor in the last line is a constant independent of j, and therefore equal to r_t itself. For $\delta \int L dt = 0$, Euler-Lagrange conditions[3] are

$$\frac{\partial L}{\partial K^i} - \frac{\frac{d}{dt} \frac{\partial L}{\partial K^i}}{\partial \dot{K}^i} = 0 = \lambda \frac{\partial F}{\partial K^i} - \dot{\lambda} \frac{\partial F}{\partial \dot{K}^i} - \lambda \frac{d}{dt} \frac{\partial F}{\partial \dot{K}^i}$$

$$(i = 1, 2, ..., n)$$

or

$$-\frac{\dot{\lambda}}{\lambda} = \frac{\frac{\partial F}{\partial K^i}}{-\frac{\partial F}{\partial K^i}} + \frac{\frac{d}{dt} \frac{\partial F}{\partial K^i}}{\frac{\partial F}{\partial K^i}}$$

$$= \frac{R^i}{P^i} + \frac{P^i_t}{P^i} \qquad (i = 1, ..., n) \quad (6)$$

$$= r_t, \quad \text{if the last relation of (5) is to hold.}$$

Thus, from the definition of W_t and r_t and (6), we have

$$W_t = W_0 \int_0^t \exp r_s \, ds = \frac{W_0 \lambda_0}{\lambda_t} \qquad (7)$$

But, recalling the definition of P^i and Y in (5), our energy integral also takes the form

$$\lambda_t Y_t \equiv \lambda_0 Y_0 \qquad (4')$$

Combining (4′) and (7), we arrive at our fundamental economic law of conservation of the capital-output ratio, $W_t / Y_t \equiv W_0 / Y_0$, which I have deduced elsewhere[4] as the energy integral of an equivalent unconstrained minimum-time problem.

For efficient balanced growth, or for the unique path through any initial K_0 that approaches this von Neumann turnpike asymptotically, the constant ratio will be the reciprocal of the von Neumann interest rate or of the equivalent maximal rate of balanced growth, $g = r^*$.

Notes

1. E. Whittaker, *Analytical Dynamics* (Cambridge, UK: Cambridge University Press, 1937).
2. R. Dorfman, P. Samuelson, and R. Solow,. *Linear Programming and Economic Analysis* (New York: McGraw-Hill, 1958).
3. P. Samuelson, *Collected Scientific Papers, I* (Cambridge, MA: MIT Press, 1965; Chs. 17 [1937], 20 [1959]).
4. P. Samuelson, "Two Conservation Laws in Theoretical Economics," Chapter 3 of this volume, in which it is shown that the present-discounted value of both capital and income are time-invariants for any efficient program in a closed von Neumann model.

Two Conservation Laws in Theoretical Economics[*]

Paul A. Samuelson

1. SUMMARY

A released body falling to the earth travels a distance proportional to the second rather than first power of its terminal velocity. This fact, belatedly recognized by Galileo, denied by Descartes, established by Huygens and Newton, culminated in the law of conservation of (mechanical) energy. Mathematically, this law is based upon a property of the calculus of variations according to which the optimizing solution to

$$O = \delta \int L(q, \dot{q})\, dt = \delta \int [\Sigma\, \Sigma\, a_{ij}\, (q_1, \ldots)\, \dot{q}_i\, \dot{q}_j - V\, (q_1, \ldots)]\, dt$$

has, because the integrand does not involve time explicitly, an "energy" integral of the form

$$L - \sum_{i=1}^{n} \dot{q}_i\, \partial L\, /\, \partial \dot{q}_i = \text{constant} = \text{kinetic energy} + \text{potential energy.}[1]$$

For a closed, consumptionless system of von Neumann type,[2] the net capital formations of a vector of capital goods, $(d\,K_t^j\, /\, dt) = (\dot{K}_t^j)$ $= \dot{K}_j$ are assumed to be related to the current rate of production of a

---

[*]I owe thanks to the National Science Foundation. In Chapter 2, I prove similar relations without considering the minimum-time formulations.

consumption good C_t, and to the vector of stocks of those capitals, K_t $= K_t^j$ by a smooth, neoclassical, concave, first-degree-homogeneous transformation function, $C_t = F(K_t ; \dot{K}_t) \equiv 0$ in a closed von Neumann system. To ensure that the system, beginning with vectoral initial capital K_0, will reach any terminal capital K_t in minimum elapsed time requires solving an optimal control problem of variational type. Two economic conservation laws are here derived for such a system.

One states the formal truth that the value of net output (net national product or income), expressed in terms of any selected good as *numeraire* , accumulates through time at the (*numeraire*-good's own) rate of interest—just as the value of capital does.

The second, and related conservation law, states the following invariance: *Along any optimal path, the ratio between the value of income and the value of capital wealth is a constant.* This invariance of the (aggregate) capital-output ratio would not be remarkable in balanced-growth configurations where there are no relative price changes and capital gains; but the present result holds for *any* optimal program, whatever the changes in relative prices and factor proportions. That this law is not trivially true is shown in the final section on discrete-time systems, in which no such conservation law is exactly valid.

2. INTERTEMPORAL EFFICIENCY[3]

The optimal-control problem is defined as

$$\underset{\{K_t^j\}}{\text{Max}} \quad \int_0^T \dot{K}^1 \, dt \tag{1}$$

subject to $0 = F (K_t^1, ..., K_t^n, \dot{K}_t^1, ..., \dot{K}_t^n)$, $(K_0^1, K_0^2, ..., K_0^n)$ and $(K_T^2, ..., K_T^n)$ prescribed.

The Euler necessary conditions for optimality are derived by the Lagrangean-multiplier technique from

$$\delta \int_0^T [\dot{K}^1 + \lambda_t\ F(K^1, \ldots; \ldots, \dot{K}^n)]\ dt = 0$$

to yield, for $i = 1, 2, \ldots, n$

$$0 = \frac{d}{dt}\ (\lambda\ F_{n+i}) - \lambda\ F_i = \dot{\lambda}\ F_{n+i} + \lambda(\dot{F}_{n+i} - F_i)$$

$$0 = F \tag{2}$$

Here $\qquad F_i = \dfrac{\partial F}{\partial K^i}\ , \qquad F_{n+i} = \dfrac{\partial F}{\partial \dot{K}^i}$

Economically,[4]

$$-\frac{\dot{\lambda}_t}{\lambda_t} = \frac{-F_i}{F_{n+i}} + \frac{-\dot{F}_{n+i}}{-F_{n+i}} = \frac{R^i}{P^i} + \frac{\dot{P}^i}{P^i} = r_t \tag{3}$$

where

$i \quad = 1, 2, \ldots, n$

$R_t^i = \partial F / \partial K^i$ is the (net) rental of the ith capital

$P_t^i = -\partial F / \partial \dot{K}^i$ is the (cost) price of the ith capital

$r_t \quad$ = the (common, equalized) rate of interest of all capitals, reckoned by adding to anyone's percentage rent yield its foreseeable algebraic percentage capital gain all expressed in the numeraire units of F itself, which can be interpreted as the consumption good in the relation $C = F(K^1, \ldots; \ldots, \dot{K}^n)$. If the own-rate-of-interest in terms of the ith capital good is written as r_t^i, then equation (3) says that for $i, j = 1, 2, \ldots, n$,

$$r_t = r_t^i + \frac{\dot{P}^i}{P^i}, \qquad r_t^i = r_t^i + \left(\frac{d}{dt} \frac{P^j}{P^i} \right) / \frac{P^j}{P^i}$$

Equivalent to equation (3) are

$$\frac{\lambda_t}{\lambda_0} = \exp - \int_0^t r_s \, ds, \qquad \text{present-discounted value} \qquad (3a)$$

$$\left(\frac{\lambda_t}{\lambda_0} \right)^{-1} = \int_0^t \exp r_s \, ds, \qquad \text{compound-interest accumulation. (3b)}$$

3. LAW OF CONSERVATION OF INCOME GROWTH

We may now write down the energy integral for the Lagrangean integrand $\dot{K}^1 + \lambda F$, which depends on time only through the dependence of its variables (K_t^j, λ_t) on time. Hence

$$\dot{K}^1 + \lambda F - \sum_{j=1}^n \dot{K}^j \frac{\partial [\dot{K}^1 + \lambda_t F]}{\partial \dot{K}_j} \Big|_t \equiv \text{constant, or}$$

$$\dot{K}_t^1 - 0 - \dot{K}_t^1 + \lambda_t \sum_1^n \dot{K}_t^j (F_{n+j})_t \equiv \alpha \qquad (4)$$

Finally,

$$\sum_1^n \dot{K}_t^j (-F_{n+j})_t \equiv \frac{\lambda_0}{\lambda_t} \sum \dot{K}_0^j (-F_{n+j})_0, \text{ or} \qquad (5a)$$

$$Y_t = \sum_1^n P_t^j \dot{K}_t^j \equiv \left(\sum_1^n P_0^j \dot{K}_0^j \right) \exp \int_0^t r_s \, ds \qquad (5b)$$

$$= Y_0 \exp \int_0^t r_s \, ds \equiv Y_T \exp - \int_t^T r_s \, ds$$

where Y_t is the usual symbol for national income or product: equal to the flow of output $\sum P^j_t \dot{K}^j_t$, consumption being zero; or equal to the flow of input rents $\sum R^j_t K^j_t$ —everything reckoned in units of F as *numeraire* and both definitions adding up to the same total by virtue of Euler's theorem on homogeneous functions.

This first conservation law says: *Total income along any optimal path (reckoned in terms of **any** numeraire) grows in a closed system at the (variable) compound interest yield of the system (expressed in the same numeraire).*

If it had been total capital or wealth that grows at this compound-interest rate, one would have been tempted to say, "What is so remarkable about that? Having assumed a closed system in which consumption is zero and all 'income' is plowed back in the system as saving, how else could we expect wealth to grow?" But here it is income produced that our law is referring to, and the finding is a more subtle one.

Where wealth is concerned, F's homogeneity confirms the formalism that its total value at any time, W_t, growing at the compound interest rate will be compatible with its instantaneous yield's being at any time equal to its total percentage net rental plus the percentage rise in its cost price. Thus, applying equation (3) to the composite (K^j)

$$r_t = \frac{\sum\limits_1^n R^j_t K^j_t}{\sum\limits_1^n P^j_t K^j_t} + \frac{\sum\limits_1^n \dot{P}^j_t K^j_t}{\sum\limits_1^n P^j_t K^j_t} = \frac{\sum\limits_1^n P^j_t \dot{K}^j_t}{\sum\limits_1^n P^j_t K^j_t} + \frac{\sum\limits_1^n \dot{P}^j_t K^j_t}{\sum\limits_1^n P^j_t K^j_t}$$

$$= \frac{\dfrac{d}{dt}\sum\limits_1^n P^j_t K^j_t}{\sum\limits_1^n P^j_t K^j_t} = \frac{\dot{W}_t}{W_t}$$

$$(6)$$

But why should the middle term of the middle equation above be a constant[5] when r_t is not a constant? Or, why should the last aggregate capital-gains term in that equation move opposite to r_t in the exact degree necessary to keep the middle Y_t / W_t term constant? Yet that is what our first conservation law does require—in order that both income and wealth should each grow in the same compound-interest fashion.

I now proceed to devise the second economic conservation law, confirming that the ratio of wealth to income is a strict constant along each optimal path.

4. LAW OF CONSERVATION OF THE CAPITAL-OUTPUT RATIO

It is known that our constrained-maximum problem involving time as the independent variable can, for a closed consumptionless system, be converted into an equivalent unconstrained-minimum problem in which any capital good, say K^1, is taken as the independent variable, and where the remaining K^j are chosen to minimize time, T, needed to go from prescribed initial K_0^j to prescribed terminal K_T^j.

The original problem of (1) above now[6] becomes

$$\operatorname*{Min}_{K^j (K^1)} \int_{K_0^1}^{K_T^1} \left(\frac{dt}{dK^1} \right) dK^1 = \int_{K_0^1}^{K_T^1} \overline{L} (K^1, K^2, \ldots; \frac{dK^2}{dK^1}, \ldots \frac{dK^n}{dK^1}) \, dK^1 \quad (7)$$

where the new integrand $\overline{L}$ denotes $dt / dK^1 = (\dot{K}^1)^{-1}$ defined by solving the following implicit function for $(\dot{K}^1)^{-1}$

$$0 = F (K^1, K^2, \ldots; \dot{K}^1, \dot{K}^2, \ldots)$$

$$= F (K^1, K^2, \ldots; \dot{K}^1, \left(\frac{\dot{K}^2}{\dot{K}^1} \right) \dot{K}^1, \ldots)$$

$$= F\ (K^1, K^2, \ldots;\ \dot{K}^1, \dot{K}^1\frac{dK^2}{dK^1},\ \ldots)$$

This solution is always unique for paths where all K's grow because the Jacobian

$J = F_{n+1} + \sum_{2}^{n} F_{n+j}\ \frac{dK_j}{dK^1}$ is negative along such paths by virtue of the

fact that $-F_{n+j} = P_j > 0$.

Actually it will always be possible to transform $\bar{L}dK_1$ in the integral that is to be minimized into the form $Ld\,\log K_1$, where the new L does not anywhere involve the independent variable explicitly. And thus we can get our second conservation law in the form of an energy integral.

To postpone tedious calculations, first consider the special joint-production case

$$0 = F\ (K, \dot{K}) \equiv Y\ [K^1, \ldots, K^n] - I\ [\dot{K}^1, \ldots, \dot{K}^n] \qquad (8)$$

where Y and $-I$ are both concave, smooth and homogeneous-of-the-first degree. A well-known example[7] is

$$0 = (K^1)^\alpha\ (K^2)^{1-\alpha} - [\ 1/2(\dot{K}^1)^2 + 1/2(\dot{K}^2)^2\ ]^{1/2}$$

Economically, Y is national income and I is the equivalent flow of national product, which in the case of zero consumption consists of net investment. From

$$0 = Y\ [1,\ \frac{K^2}{K^1},\ \ldots] - I\ [1, \frac{\dot{K}^2}{\dot{K}^1},\ \ldots]\ \frac{\dot{K}^1}{K^1}$$

we have

$$\underset{K^i (K^1)}{\text{Min}} \int Y \left[1, \frac{K^2}{K^1}, \dots \right]^{-1} I \left[1, \frac{dK^2}{dK^1}, \dots \right] d \log K^1$$

$$= \underset{x_j (x_1)}{\text{Min}} \int Y \left[1, e^{x_2}, \dots \right]^{-1} I \left[1, (1 + x_2') e^{x_2}, \dots \right] dx_1$$

and where $(\log K_T^i)$ and $(\log K_0^i)$ are prescribed

$$x_1 = \log K^1, \dots, x_i = \log \frac{K^i}{K^1}, \dots$$

$$(9)$$

$$K^1 = e^{x_1}, \dots, \frac{K^i}{K^1} = e^{x_1}$$

$$x_i' = \frac{d\log K^i - d\log K^1}{d\log K^1} = \dots$$

$$(1 + x_i') e^{x_1} = \frac{dK^i}{dK^1}$$

Since L does not involve x_1 explicitly, any optimizing program must satisfy the energy integral

$$L - \sum_{i=2}^{n} x_i' \frac{\partial L}{\partial x_i'} = \text{constant, or}$$

$$Y^{-1} I - \sum_{i=2}^{n} x_i' Y^{-1} e^{x_i} I_i = \beta \qquad (10a)$$

$$I - \sum_{i=2}^{n} (1 + x_i') e^{x_i} I_i + \sum_{i=2}^{n} e^{x_i} I_i = \beta Y$$

$$\left[I \left(1, \frac{dK^2}{dK^1}, \dots \right) - \sum_2^n I_i \frac{dK^i}{dK^1} \right] + \sum_2^n I_i \frac{K^i}{K^1} = \beta Y$$

$$\sum_1^n I_i K^i = \beta Y$$

or

$$\sum_1^n \frac{P_t^i K_t^i}{Y_t} = \frac{W_t}{Y_t} \underset{t}{\equiv} \frac{W_0}{Y_0} = \frac{\sum_1^n P_0^i K_0^i}{\sum_1^n P_0^i \dot{K}_0^i} = \beta \qquad (10b)$$

Thus, the capital-output ratio is indeed constant along any optimal path, despite the occurrence of (foreseen) relative price changes and capital gains terms.   Q.E.D.

For the special case of maximal balanced exponential growth at rate g along the von Neumann turnpike, of course all P's are constant and the interest rate is a constant at the level g. The capital output ratio β is then $1/g$ in accordance with the usual rules of capitalization (e.g., $r = .05$ implies "20-years purchase").

To prove all this when F cannot be split up into $Y - V$, we solve for L in

$$0 = F(K^1, K^2, \dots; \dot{K}^1, \dot{K}^2, \dots)$$

$$= F \left[1, \frac{K^2}{K^1}, \dots; L^{-1}, L^{-1} \left(\frac{\dot{K}^2}{\dot{K}^1} \right) \left(\frac{K^1}{K^2} \right) \left(\frac{K^2}{K^1} \right), \dots \right]$$

$$= F \left[1, e^{x_2}, \dots; L^{-1}, L^{-1} (1 + x_2') e^{x_2}, \dots \right]$$

and note that

$$\frac{\partial L}{\partial x_i'} = \frac{L^{-1} F_{n+1} e^{x_i}}{(F_{n+1} + \sum_2^n F_{n+j} (1 + x_j') e^{x_j}) L^{-2}} \tag{11}$$

$$L - \sum_2^n x_i' \frac{\partial L}{\partial x_i'} = \text{constant} \tag{12a}$$

$$\frac{L - \sum_2^n x_i' \left(L F_{n+i} e^{x_i} \right)}{\left[F_{n+1} + \sum_2^n F_{n+j} \left(1 + x_j' \right) e^{x_j} \right]} F_{n+1} + \sum_2^n F_{n+j} \, e^{x_j}$$

$$+ \sum_2^n P_{n+j} \, x_j' \, e^{x_j} - \sum_2^n F_{n+i} \, x_i' \, e^{x_i}$$

$$= L^{-1} \beta \left[F_{n+1} + \sum_2^n F_{n+j} \left(1 + x_j' \right) e^{x_j} \right] F_{n+1} K^1$$

$$+ \sum_2^n F_{n+j} \, K^j + 0 = \beta \left[F_{n+1} \dot{K}^1 + \sum_2^n F_{n+j} \, \dot{K}^j \right]$$

$$\frac{\sum_1^n P_t^j K_t^j}{\sum_1^n P_t^j \dot{K}_t^j} \equiv \frac{W_t}{Y_t} \equiv^t \frac{W_0}{Y_0} \equiv \beta, \qquad \text{Q.E.D.} \tag{12b}$$

5. CONCLUSION

The two conservation laws derived here are of course not independent. From the fact that the first law says that income Y_t grows at compound interest,[8] which is the same as how wealth is defined to grow, one could infer the content of the second law, namely that the ratio of wealth to income must be invariant along an optimal path (or in every perfect competitive equilibrium in a futures market). But all this consistency depends upon the assumption of constant returns to scale (i.e., F a homogeneous-first-degree function). Thus, if F were not homogeneous, the first law would still be valid: $\sum P_t^j K_t^j$ would grow at compound interest (even if the *other* definition of income, $\sum R^j K^j$, grew in some other way; and if the present-discounted value of this stream of rents no longer equaled the cost-value of wealth, $\sum R^j K^j$). But with F not homogeneous (as in a model where labor is held constant and is limiting), L in $\partial \int L dx_1 = 0$ would depend irreducibly on x_1 explicity, and there would be no energy integral of any kind.

Will these neoclassical results hold for a neo-classical model of Joan Robinson type,[9] where marginal productivities in the form of partial derivatives are not definable because only a finite number of activities are technologically possible? I believe the answer is yes in the following sense: Along every optimal path, at least those characterized by equalities rather than inequalities, there will exist time profiles of prices and of profit rates such that income will grow at compound interest and the capital-output rate will be constant.

Finally, a glance at the discrete-time case, where $C_{t+1} = G\left(K_t; K_{t+1}\right) \equiv 0$, will show how singular the above conservation laws are, since for the discrete case no such invariance will hold. Corresponding to the calculus of variations problem

$$\underset{u_t}{\text{Max}} \int_0^T L\left(u, \dot{u}\right) dt$$

for which an energy integral always holds, we replace integral by sum
and contemplate the discrete-time problem

$$\underset{u_t}{\text{Max}} \sum_0^T S\,(u_t, u_{t+1})$$

for which no such invariance is valid. What is still valid is the Euler-like variational condition which involves the second-order difference equation

$$0 = \frac{\partial S\,(u_t, u_{t+1})}{\partial u_t} + \frac{\partial S\,(u_{t-1}, u_t)}{\partial u_t}$$

$$(t = 1, 2, ..., T)$$

Economically, for $n = 2$, our von Neumann system is defined by

$$\text{Max } K^1_{T+1}$$

subject to $0 = G\,(K_t\,; K_{t+1}), \quad (K^1_0, K^2_0)$ and K^2_{T+1} prescribed, $\quad T \geq 1$

The necessary Euler condition is of familiar envelope type[10]

$$\frac{\dfrac{\partial G\,(K^1_t, ...\,)}{\partial K^j_t}}{\dfrac{\partial G\,(K^1_t, ...\,)}{\partial K^1_t}} = \frac{\dfrac{\partial G\,(K^1_{t-1}, ...\,)}{\partial K^j_t}}{\dfrac{\partial G\,(K^1_{t-1}, ...\,)}{\partial K^1_t}} = \frac{P^j_t}{P^i_t}$$

To show there is no invariance of the ratio $W_t\,/\,Y_t$ or $W_{t+1}\,/\,Y_t$ type, it will suffice to consider any one simple example with $T = 1$. Thus, let

$$G\,(K_t\,; K_{t+1}) = K^1_t + K^2_t - [\,(K^1_{t+1} - K^1_t\,)^2 + (K^2_{t+1} - K^2_t\,)^2\,]^{1/2} = 0$$

This G, which is of the type cited in Footnote 7, is concave in the K's and homogeneous of the first degree.

It will suffice to consider any arbitrary efficient path as for example

$$(K_0^1, K_0^2; K_1^1, K_1^2, K_2^2) = (3, 2; 6, 6; \quad 322/25, \quad 204/25),$$

with

$(P_0^1, P_0^2; P_1^1, P_1^2, P_2^1; P_2^2)$ calculable as

$(60/150, 30/150; 90/150, 120/150; 86/150, 27/150)$,

(W_0, W_1, W_2) as $(120/75, 630/75, 664/75)$ and

(Y_0, Y_1, Y_2) as $(3 - 5, 2, 13/9)$.

Clearly, there is no way of forming ratios of the above W and Y numbers so as to preserve constancy over time.

I conclude that, at best, if discrete-time intervals can be taken so small that there is little error in approximating the system by a continuous-time model, we can hope to state an "almost-constancy" of the capital-output ratio along any optimal path.

Notes

1. See any treatise on classical mechanics, such as E. T. Whittaker, (1937), *Analytical Dynamics,* Chapter III, p. 62.
2. The differentiable, neoclassical version of the famous von Neumann system is discussed in R. Dorfman, P. Samuelson, and R. Solow (1958), *Linear Programming and Economic Analysis,* New York: McGraw-Hill, Chapters 11 and 12.

3. See P. A. Samuelson, "Efficient Paths of Capital Accumulation in Terms of the Calculus of Variations," pp. 77-88 in K. J. Arrow, S. Karlin, and P. Suppes, eds. (1960), *Mathematical Methods in the Social Sciences, 1959*, Stanford, California: Stanford University Press. This is reproduced as Chapter 26 in Samuelson's (1965) *Collected Scientific Papers, Vol. I*, J. E. Stiglitz, ed., Cambridge, Massachusetts: MIT Press.

4. This efficiency condition is derived in P. A. Samuelson, op. cit., p. 79 in slightly different notation. As arbitrage conditions for own-interest rates in a perfect capital market, they were derived in P. A. Samuelson (1937), "Some Aspects of the Pure Theory of Capital," *Quarterly Journal of Economics, LI*, pp. 469-496, particularly on p. 490 of pp. 485-491. (This is reproduced as Chapter 17 in *Collected Scientific Papers, Vol. I*, op. cit.)

5. If we had but one capital good, F would take the simple form $0 = rK\text{-}\dot{K}$; with *r* a constant, the problem would become trivial, since rK would grow at the same exponential rate as K. (The reader is warned that Irving Fisher would not call $Y = \sum F_j K_j = S(-F_{n+j}) K_j$ "income," reserving that word for "consumption," which is zero in the present closed system. But this is the usual Marshall-Haig definition of income. If, with Mrs. Robinson we call $\sum P^j K^j$ a "Wicksell effect," it follows that such effects must behave in the described way along any optimal path.

6. See p. 85 of the 1960 Samuelson paper cited in Footnote 3.

7. This case is utilized in C. Caton and K. Shell (1971), "An Exercise in the Theory of Heterogeneous Capital," *Review of Economic Studies, 38*, pp. 13–22. It was suggested earlier by P. A. Samuelson (1966), "The Fundamental Singularity Theorem for Non-Joint Production," *International Economic Review, VII*, pp. 34–41, as have E. Burmeister and various collaborators.

8. Another way of formulating the present finding is to say that "the present-discounted value of any future-time investment or income is a strict constant, just as the present-discounted value of any future-time capital stock is a (different) constant. As James Mirrlees states in private correspondence, it is the truth of the first of these that is at first surprising. Edwin Burmeister has also pointed out to me that the constancy of a Pontryagin Hamiltonian is another way of expressing and proving the present result, which is reasonable in view of the classical fact from mechanics that Hamilton's canonical $H(p, q)$ is given the interpretation of total energy.

9. For such a Robinson case, see M. Bruno (1967), "Optimal Accumulation in Discrete Capital Models," in K. Shell, ed., *Essays on the Theory of Optimal Economic Growth*, Cambridge, Massachusetts: MIT Press, Essay II, pp. 181-218. (However, the closed von Neumann case is not dealt with by Dr. Bruno.)

10. See Dorman, Samuelson, and Solow, op. cit., pp. 312–313.

CHAPTER 4

The Invariance Principle and Income-Wealth Conservation Laws

*Ryuzo Sato**

1. INTRODUCTION

In the early part of the 19th century William Rowan Hamilton discovered a principle which can be generalized to encompass many areas of physics, engineering and applied mathematics. Hamilton's principle roughly states that the evolution in time of a dynamic system takes place in such a manner that integral of the difference between the kinetic and potential energies for the system is stationary. If the 'action' integral is free of the time variable, the sum of the kinetic and potential energies, the Hamiltonian, is constant—the conservation law of the total energy.

In the early part of this century Emmy Noether (1918) discovered the fundamental invariance principle known as the *Noether Theorem*.[1] This principle not only extended the idea of Hamilton and the conservation law of total energy, but also provided the formal methodology to study the general 'invariance' problem of a dynamic system. Influenced by the work of Klein (1918) and of Lie (1891) on the transformation properties of differential equations under continuous (Lie) groups, Noether had the ingenious insight of combining the

*The author wishes to express his appreciation to Paul A. Samuelson, William A. Barnett, Hal R. Varian, Gilbert Suzawa, Takayuki Nôno, Fumitake Mimura, and Shigeru Maeda for their helpful comments on an earlier version of this chaper.

methods of the formal calculus of variation with those of Lie group theory. Since the first application of the Noether invariance principle to particle mechanics by Bessel-Hagan (1921), this new area of mathematics has exhibited remarkable development in the last fifty years.[2]

The study of economic conservation laws is still in its infancy compared with its counterparts in physics and engineering. Yet this is an area where rapid progress is being made. In economics the conservation law has its roots in Ramsey (1928). But it was Samuelson (1970) who first explicitly introduced conservation laws to theoretical economics. The recent work by Weitzman (1976), Sato (1981, 1982), Kemp and Long (1982), Samuelson (1982), Sato, Nôno, and Mimura (1983) and Kataoka (1983) provide an indication of the rapid progress being made in this field and the great interest shown in the analysis of economic conservation laws. In Ramsey (1928), Samuelson (1970, 1982), Weitzman (1976) and Kemp and Long (1982), the authors used the standard invariance condition of the calculus of variation, which is a special case of the Noether invariance principle, while in Sato (1981, 1982) and Sato, Nôno and Mimura (1983) we employed the Noether invariance principle and, thus, were able to obtain the more general results of 'hidden' conservation laws.

The purpose of this paper is to extend the work on economic conservation laws in the direction of uncovering more 'hidden' conservation laws. We shall briefly discuss the application of the conservation laws to the econometric estimation of the optimal growth models. However, before we do so, we shall begin with a brief summary of the literature.

2. BRIEF SUMMARY OF THE LITERATURE

Ramsey's (1928, p. 547) famous rule of optimal saving is the first reference (implicit) to a conservation law in economics. His rule states that

$$\dot{c} = \frac{dc}{dt} = f(a, c) - x = \frac{B - (U(x) - V(a))}{u(x)} \qquad (2.1)$$

where f = production function, a = labor, c = capital, x = consumption, $U(x)$ = utility of consumption, $V(a)$ = disutility of labor, $u(x) = \partial U / \partial x$ = marginal utility of consumption. This rule is an *economic version of the law of conservation of energy*. Letting L (Lagrangean) = $-(U(x) - V(a))$ and Bliss $B = H$ (Hamiltonian), the Ramsey rule is derived from the conservation law,

$$H = B = u(x) \, \dot{c} + (U(x) - V(a)) = u(x) \, \dot{c} - L \qquad (2.2)$$

Samuelson (1970a) is the first economist to *explicitly* introduce conservation laws in theoretical economics. From the analogy of the law of conservation of (physical) energy, kinetic energy + potential energy = constant, Samuelson obtained the *conservation law of the aggregate capital-output ratio* in a neoclassical von Neumann economy, where all output is saved to provide capital formation for the system's growth.

A typical optimization model in modern economics deals with the problem of maximizing

$$\int_0^T L(t, k, \dot{k}) \, dt = \int_0^T D(t) \, U \, [f(t, k, \dot{k})] dt, \qquad (2.3)$$

where $D(t)$ = discount rate factor, U = utility function, f = consumption possibility function, k = capital-labor ratio and $\dot{k}$ = time derivative of k. Here, we also assume that U and f are neoclassical functions satisfying all the regularity conditions. In his recent work, Weitzman (1976) has derived the invariance condition of the income-wealth ratio in a neoclassical model (2.3) when the welfare of the society is discounted at a constant rate for $T \to \infty$ and f is free of technical progress. A similar result is obtained by such authors as Kemp and Long (1982), Samuelson (1982) and Sato (1982).

The first derivation of economic conservation laws via the application of Noether's Theorem and Lie groups is contained in Sato (1981), where some of the existing known results are confirmed, and also several new hidden laws are uncovered. For instance, it is shown (Sato, 1981, p. 279) that Samuelson's (1970b) two conservation laws are the *only* laws globally operating for the von Neumann system of optimal growth. Also in studying the Ramsey and general neoclassical growth models, it was discovered that there exist several new conservation laws operating in the vicinity of the steady state. For example, in the Liviatan-Samuelson model (1969), *the discounted welfare measured in terms of the modified income, which is the sum of production income and rental income is constant* [Sato (1981, p. 264)]. The problem of how various types of technical progress will affect the conservation laws was studied by Sato, Nôno and Mimura (1983). By redefining income and supply price of capital, we were able to uncover several 'hidden' conservation laws.

In earlier works [e.g. Sato (1981)], an attempt was made to discover conservation laws which are *independent of the forms of production and utility functions*. In the present paper we intend to incorporate these functions in conservation laws, which will enable us to uncover more hidden invariances.

3. A MODEL WITH HETEROGENEOUS CAPITAL GOODS

Let utility depend on consumption of many goods which in turn depend on a vector of capital goods, $k = (k^1,..., k^n)$, and a vector of investment, $\dot{k} = (\dot{k}^1,..., \dot{k}^n)$, so that

$$U = U(\dot{k}, k) \tag{3.1}$$

Let $U(\dot{k}, k)$ be a strictly concave function with existent partial derivatives, for which

$$U\,(0,\,k) < U\,(0,\,\bar{k}\,) = 0 \qquad k \neq \bar{k},$$

$$U_i(0,\,\bar{k}\,) \neq 0, \qquad U_i = \partial U\,/\,\partial k^{\,i}, \qquad i = 1,\,...,\,n. \tag{3.2}$$

The society's problem is to maximize the welfare functional

$$J = \int_0^\infty e^{-\rho t}\,U[\dot{k}\,(t),\,k\,(t)]\,dt, \qquad \rho > 0, \tag{3.3}$$

subject to the appropriate initial conditions. For simplicity we shall write

$$L\,(t,\,k,\,\dot{k}) = e^{-\rho t}\,U\,(\dot{k}\,(t),\,k\,(t)). \tag{3.4}$$

The necessary condition for the optimal solution is that the Euler-Lagrange equations vanish:

$$E_i = \frac{\partial L}{\partial k^{\,i}} - \frac{d}{dt}\left(\frac{\partial L}{\partial \dot{k}^{\,i}}\right) = e^{-\rho t}\,\frac{\partial U}{\partial k^{\,i}} - \frac{d}{dt}\left(e^{-\rho t}\,\frac{\partial U}{\partial \dot{k}^{\,i}}\right)$$

$$\tag{3.5a}$$

$$= e^{-\rho t}\left[\frac{\partial U}{\partial k^{\,i}} + \rho\,\frac{\partial U}{\partial \dot{k}^{\,i}} - \frac{d}{dt}\left(\frac{\partial U}{\partial \dot{k}^{\,i}}\right)\right] = 0$$

which implies that

$$E_i = 0 \Leftrightarrow \frac{\partial U}{\partial k^{\,i}} + \rho\,\frac{\partial U}{\partial \dot{k}^{\,i}} - \frac{d}{dt}\left(\frac{\partial U}{\partial \dot{k}^{\,i}}\right) = 0, \quad i = 1,\,...,\,n. \tag{3.5b}$$

If we define the supply price of the ith capital as $p^i = -\,\partial U\,/\,\partial \dot{k}^{\,i}$, (3.5) states that the time derivative of $p^{\,i}$ is equal to

$$\dot{p}^{\,i} = -\,\partial U\,/\,\partial k^i + \rho p^i, \qquad i = 1,\,...,\,n. \tag{3.5c}$$

4. NOETHER'S THEOREM (INVARIANCE PRINCIPLE)[3]

Before we present the analysis of economic conservation laws (hidden or unhidden) associated with the above model, we briefly discuss the mathematical properties of Noether's theorem and invariance identities. As this theorem is relatively unknown to economists, we shall begin with the definitions of dynamic symmetry (or invariance).

Let us consider a Lagrange function L, which is twice continuously differentiable in each of its $2n + 1$ arguments. We then have the variational integral

$$J(x) = \int_a^b L(t, x(t), \dot{x}(t)) \, dt, \qquad (4.1)$$

where $x =$ the set of all vector functions $x(t) = (x^1(t), \ldots, x^n(t))$, $t \in [a, b]$. The type of invariance transformations that will be considered are (technical change) transformations of configuration space, i.e., $(t, x^1, \ldots, x^n)$ – space, which depend upon r real, independent (essential) parameters $\varepsilon^1, \ldots, \varepsilon^r$. To be more precise, we require here that the transformations are given by

$$\bar{t} = \phi(t, x, \varepsilon), \qquad \varepsilon = (\varepsilon^1, \ldots, \varepsilon^r),$$

$$x^i = \psi^i(t, x, \varepsilon), \qquad i = 1, \ldots, n. \qquad (4.2)$$

In economic terms, ϕ can be considered as 'subjective' time [Samuelson (1976)], while ψ^i represent 'technical' or 'taste' change. We assume that

$$\phi(t, x, 0) = t,$$

$$\psi^i(t, x, 0) = x^i, \qquad i = 1, \ldots, n. \qquad (4.3)$$

Expanding the right-hand sides of (4.2) in Taylor series around $\varepsilon = 0$, we obtain

$$\bar{t} = t + \tau_s(t, x)\, \varepsilon^s + O\,(\varepsilon)$$

$$\bar{x}^i = x^i + \xi_s^i(t, x)\, \varepsilon^s + O\,(\varepsilon), \quad i = 1, \ldots, n. \tag{4.4}$$

$s = 1, \ldots, r,$ summation convention in force.[4]

The principal linear parts τ_s and ξ_s^i are called the *infinitesimal generators* (or *transformations*) of the transformational (4.2) given by[5]

$$\tau_s(t, x) = \frac{\partial \phi}{\partial \varepsilon^s}\,(t, x, 0), \quad \xi_s^i(t, x) = \frac{\partial \psi^i}{\partial \varepsilon^s}\,(t, x, 0) \tag{4.5a}$$

Using the customary symbol of the infinitesimal transformations, we write (4.5a) as

$$X_s = \tau_s(t, x)\,\frac{\partial}{\partial t} + \xi_s^i(t, x)\,\frac{\partial}{\partial x^i} + \left(\frac{d\xi_s^i}{dt} - x'\,\frac{d\tau_s}{dt}\right)\frac{\partial}{\partial \dot{x}^i} \tag{4.5b}$$

Definition 1. *(Dynamic invariance or dynamic symmetry)*.
(a) The fundamental integral (4.1) is absolutely invariant under the r-parameter family of transformations (4.2) if and only if we have

$$\int_{\bar{t}_1}^{\bar{t}_2} L\left(\bar{t}, \bar{x}(\bar{t}), \frac{d\bar{x}(\bar{t})}{d\bar{t}}\right) d\bar{t} - \int_{t_1}^{t_2} L\left(t, x(t), \frac{dx(t)}{dt}\right) dt = O(\varepsilon) \tag{4.6}$$

(b) Alternatively, the fundamental integral is absolutely invariant if and only if

$$L\left(\bar{t}, \bar{x}(\bar{t}), \frac{d\bar{x}(\bar{t})}{d\bar{t}}\right)\frac{d\bar{t}}{dt} - L\left(t, x(t), \frac{dx(t)}{dt}\right) = O\,(\varepsilon). \tag{4.7a}$$

Definition 2 *(Divergence-invariance or invariance with nullity).*
The fundamental integral is divergence-invariant or invariant up to a
divergence term, if there exist r functions Φ_s such that

$$L\left(\bar{t}, \bar{x}(\bar{t}), \frac{d\bar{x}(\bar{t})}{d\bar{t}} \right) \frac{d\bar{t}}{dt} - L\left(t, x(t), x(t) \right)$$

(4.7b)

$$= \varepsilon^s \frac{d\Phi_s}{dt} (t, x(t)) + O(\varepsilon), \text{ summation convention on } s,$$

with the remaining conditions of Definition 1 holding true.

This definition is more general than Definition 1 in that the left-
hand side of (4.7b) is equal to the sum of the exact differentials
multiplied by the essential parameters and to first-order terms in ε.
Since the addition of the total exact differentials does *not* change the
original Euler-Langrange equations associated with L [see Sagan
(1969)], we may call this case the *dynamic symmetry condition with
nullity*, or simply the *dynamic invariance with nullity*. Alternatively
Definition 2 is referred to as 'invariance up to an exact differential.'
[See Rund (1966, p.73)]. Then the *fundamental invariance identities*
are given by

$$X_s L + L \frac{d\tau_s}{dt} = \frac{d\Phi_s}{dt},$$

(4.8a)

or

$$\frac{\partial L}{\partial t} \tau_s + \frac{\partial L}{\partial x^i} \xi_s^i + \frac{\partial L}{\partial \dot{x}^i} \left(\frac{d\xi_s^i}{dt} - \dot{x}^i \frac{d\tau_s}{dt} \right) + L \frac{d\tau_s}{dt} = \frac{d\Phi_s}{dt},$$

(4.8b)

$$s = 1, \ldots, r, \text{ summation on } i.$$

[See Sato (1981, p. 244, eq. 17) for the derivation of (4.8).]

If the fundamental integral is invariant (either absolutely or up to an exact differential) under an r-parameter family (or group) of continuous transformations, the corresponding Lagrangean must satisfy certain conditions involving the Lagrangean, its derivatives and the infinitesimal generators of the continuous transformations. Using the Euler-Lagrange expressions the resulting transformation of these conditions is usually referred to as Noether's theorem. The importance of this theorem lies in the fact that it allows one to construct quantities which are constant along any extremal, that is, a curve which satisfies the Euler-Lagrange equations. Hence, it allows one to obtain relations which may be interpreted as 'conservation laws.'

We can now present [Sato (1981)]:

Noether's Invariance Theorem. If the fundamental integral of a problem in the calculus of variations is divergence-invariant under the r-parameter family of transformations, then r distinct quantities Ω_s (s = 1, ..., r) are constant along any extremal.

In physical and economic applications, those quantities are thus interpreted as the 'conservation laws' of the system, where

$$\Omega_s = -H\tau_s + \frac{\partial L}{\partial x^i}\,\xi_s^i - \Phi_s = \text{constant}, \quad s = 1, \ldots, r. \qquad (4.9a)$$

where H is the Hamiltonian defined by

$$H = -L + \dot{x}^i \frac{\partial L}{\partial \dot{x}^i}. \qquad (4.9b)$$

[See Sato (1981, pp. 242-251) for proof.]

5. INCOME-WEALTH CONSERVATION LAWS

Using Noether's invariance principle we are now in a position to study the dynamic symmetry conditions in the heterogeneous capital model. Here we require that t and k^i are subjected to the transformations

$$\bar{t} = \phi(t, k, \varepsilon), \qquad \varepsilon = (\varepsilon^1, \dots, \varepsilon^r),$$

$$\bar{k}^i = \psi^i(t, k, \varepsilon), \qquad i = 1, \dots, n, \tag{5.1a}$$

and, hence, the infinitesimal transformations are given by

$$X_s = \tau_s(t, k)\frac{\partial}{\partial t} + \xi_s^i(t, k)\frac{\partial}{\partial k^i} + \left(\frac{d\xi_s^i}{dt} - \dot{k}^i\frac{d\tau_s}{dt}\right)\frac{\partial}{\partial k^i}. \tag{5.1b}$$

We are interested in finding the dynamic symmetry conditions

$$\int_{\bar{t}_1}^{\bar{t}_2} e^{-\rho \bar{t}}\, U\left[\bar{k}(\bar{t}), \frac{dk(\bar{t})}{dt}\right]d\bar{t} - \int_{t_1}^{t_2} e^{-\rho t}\, U\left[k(t), \frac{dk(t)}{dt}\right]dt$$

$$= \varepsilon^s\, \Phi_s(t, k(t)) + O(\varepsilon). \tag{5.2}$$

By applying the infinitesimal transformation (5.1b) on (3.4) and using the fundamental invariance identities (4.8), we obtain

$$e^{-\rho t}\left[-\rho U\tau + \frac{\partial U}{\partial k^i}\xi^i + \frac{\partial U}{\partial k^i}\left(\frac{d\xi^i}{dt} - k^i\frac{d\tau}{dt}\right) + U\frac{d\tau}{dt}\right] = \frac{d\Phi}{dt}, \tag{5.3a}$$

or

$$\left(U - \dot{k}^i\frac{\partial U}{\partial k^i}\right)\frac{d\tau}{dt} = \rho U\tau - \frac{\partial U}{\partial k^i}\xi^i - \frac{\partial U}{\partial k^i}\frac{d\xi^i}{dt} + e^{\rho t}\frac{d\Phi}{dt}. \tag{5.3b}$$

Since we assume that we are dealing with a one-parameter transformation group, i.e., $s = 1$, we omit the subscript 1 from τ_1, ξ_1^i, etc.

Warning:

$$\dot{k}^i \frac{\partial U}{\partial \dot{k}^i} \quad \text{means} \quad \sum_{i=1}^{n} \dot{k}^i \frac{\partial U}{\partial \dot{k}^i},$$

$$\frac{\partial U}{\partial k^i} \xi^i \quad \text{means} \quad \sum_{i=1}^{n} \frac{\partial U}{\partial k^i} \xi^i, \text{etc.}$$

From the invariance principle (4.9a) the conservation law is now derived as

$$\Omega = \left(e^{-\rho t} U - e^{-\rho t} \dot{k}^i \frac{\partial U}{\partial \dot{k}^i} \right) \tau + e^{-\rho t} \frac{\partial U}{\partial k^i} \xi^i - \Phi = \text{constant}, \quad (5.4a)$$

or

$$\frac{d\Omega}{dt} = -\rho e^{-\rho t} \left(U - \dot{k}^i \frac{\partial U}{\partial \dot{k}^i} \right) \tau + e^{-\rho t} \frac{d}{dt} \left(U - \dot{k}^i \frac{\partial U}{\partial \dot{k}^i} \right) \tau$$

$$+ e^{-\rho t} \left(U - \dot{k}^i \frac{\partial U}{\partial \dot{k}^i} \right) \frac{d\tau}{dt} - \rho e^{-\rho t} \frac{\partial U}{\partial k^i} \xi^i$$

$$+ e^{-\rho t} \frac{d}{dt} \left(\frac{\partial U}{\partial k^i} \xi^i \right) - \frac{d\Phi}{dt} = 0. \quad (5.4b)$$

Hence we have

$$\left[\frac{d}{dt} \left(U - \dot{k}^i \frac{\partial U}{\partial \dot{k}^i} \right) + \rho \, \dot{k}^i \frac{\partial U}{\partial \dot{k}^i} \right] \tau$$

$$= \rho \, U\tau - \frac{\partial U}{\partial \dot{k}^i} \xi^i - \frac{\partial U}{\partial \dot{k}^i} \frac{d\xi^i}{dt} - \left(U - \dot{k}^i \frac{\partial U}{\partial \dot{k}^i} \right) \frac{d\tau}{dt}$$

$$+ e^{\rho t} \frac{d\Phi}{dt} + \left[\frac{\partial U}{\partial k^i} + \rho \frac{\partial U}{\partial \dot{k}^i} - \frac{d}{dt} \left(\frac{\partial U}{\partial \dot{k}^i} \right) \right] \xi^i. \qquad (5.4c)$$

The term inside the bracket of the last expression of the right-hand side is zero because of the vanishing condition placed on the Euler-Lagrange equantions (3.5b). Eliminating $d\Phi / dt$ between (5.3b) and (5.4c) we obtain our first conservation law

$$\frac{d}{dt} \left(U - \dot{k}^i \frac{\partial U}{\partial \dot{k}^i} \right) = - \rho \dot{k}^i \frac{\partial U}{\partial \dot{k}^i} \, .$$

It is easy to identify $U - \dot{k}^i (\partial U / \partial \dot{k}^i)$ as income in utility terms—Marshall-Haig-Kuznets' definition, while

$$- \dot{k}^i \frac{\partial U}{\partial \dot{k}^i} \equiv - \sum_{i=1}^{n} \dot{k}^i \frac{\partial U}{\partial \dot{k}^i} \equiv \sum_{i=1}^{n} p^i \dot{k}^i \qquad (5.5a)$$

is the utility-value-of-investment. Hence, this conservation law states that along the optimal path

$$\frac{d}{dt} \, (\text{income at } t) \; = \; \rho \; \times \; \left(\begin{array}{l} \text{utility-value-of-} \\ \text{investment at } t \end{array} \right) \qquad (5.5b)$$

$$\left(\begin{array}{c} \text{rate of change in} \\ \\ \text{income at } t \end{array} \right) = \rho \times \left(\begin{array}{c} \text{utility-value-of-} \\ \\ \text{investment at } t \end{array} \right) \qquad (5.5c)$$

For $\rho > 0$, we have along the optimal path

$$\frac{\dfrac{d}{dt} (\text{income at } t)}{(\text{utility-value-of-investment at } t)} = \rho = \text{constant, for all } t \quad (5.5d)$$

Our first conservation law is equivalent to what Samuelson (1982) calls a *pseudo-net productivity relation*. The *genuine net-productivity* relation for $i = 1$ can be obtained from

$$y(t) = c(t) + \dot{k}(t) = f(k(t)), \qquad (5.6a)$$

$$\frac{dy(t)}{dt} = f'(k(t))\,\dot{k}(t), \qquad (5.6b)$$

$$\dot{y} = r(t)\,\dot{k}(t), \ \lim_{t \to \infty} r(t) = \rho. \qquad (5.6c)$$

Hence, only in the limit, as $t \to \infty$, will the system settle down to stationary equilibrium at market rate equal to the ρ time-preference parmeter.

It should be noted that the utility-value-of-net-capital formation is not equivalent to the time derivative of the utility-value-of-capital. The latter usually includes capital gains, or price changes, while the former does not. The ratio of the utility-value-of-income to the utility-value-of-capital-goods is constant in the present model. We shall later discuss this aspect of the invariance conditions in a von Neumann model of economic growth (section 8).

Next we can derive the income-wealth conservation law. Let us assume that

$$\tau = 1, \tag{5.7a}$$

$$\xi^i = 0, \quad i = 1, \ldots, n. \tag{5.7b}$$

We then get, from (5.3a) and 3.5b), along the optimal path

$$\frac{d\Phi}{dt} = -\rho e^{-\rho t} U. \tag{5.8a}$$

Also from (5.4b) we have

$$\frac{d\Phi}{dt} = -\rho e^{-\rho t} \left(U - \dot{k}^i \frac{\partial U}{\partial \dot{k}^i} \right) + e^{-\rho t} \frac{d}{dt} \left(U - \dot{k}^i \frac{\partial U}{\partial \dot{k}^i} \right)$$
$$\tag{5.8b}$$
$$= \frac{d}{dt} \left[e^{-\rho t} \left(U - \dot{k}^i \frac{\partial U}{\partial \dot{k}^i} \right) \right].$$

Equating (5.8b) with (5.8a) we obtain

$$\frac{d\Phi}{dt} = \frac{d}{dt} \left[e^{-\rho t} \left(U - \dot{k}^i \frac{\partial U}{\partial \dot{k}^i} \right) \right] = -\rho e^{-\rho t} U, \tag{5.8c}$$
$$0 \leq t \leq \infty.$$

Note that the left-hand side of (5.8c) is equal to

$$\frac{d}{dt} \left[e^{-\rho t} \left(U - \dot{k}^i \frac{\partial U}{\partial \dot{k}^i} \right) \right] = -\frac{dH}{dt}, \tag{5.8d}$$

while the right-hand side of (5.8c) is equal to

$$-\rho e^{-\rho t} U = \frac{\partial L}{\partial t} \, . \tag{5.8e}$$

Hence (5.8) shows that

$$\frac{d\Phi}{dt} = \frac{\partial L}{\partial t} = -\frac{dH}{dt} \, . \tag{5.8f}$$

The null term Φ now turns out to be equal to the integral of the negative value of the Hamiltonian and to the integral of the Lagrangean function. Alternatively, we can say that the absolute value of the rate of change in the Hamiltonian must be equal to the rate of change in the null term for the entire planning period. Integrating (5.8c), we get the null term as

$$\Phi \Big|_{t}^{\infty} = e^{-\rho s} \left[U(s) - \dot{k}^i(s) \, \frac{\partial U(s)}{\partial \dot{k}^i(s)} \right] \Big|_{t}^{\infty} = -\rho \int_{t}^{\infty} e^{-\rho s} \, U(s) \, ds, \tag{5.9a}$$

or

$$\Phi \Big|_{\infty}^{t} = \rho \int_{t}^{\infty} e^{-\rho s} \, U(s) \, ds. \tag{5.9b}$$

By the concavity assumption on U and by the transversality condition, we have

$$e^{-\rho s} \left[U(s) - \dot{k}^i(s) \, \frac{\partial U(s)}{\partial \dot{k}^i(s)} \right] \to 0 \quad \text{as} \quad s \to \infty. \tag{5.10}$$

Hence, (5.9a) becomes

$$- e^{-\rho s} \left[U(s) - \dot{k}^i(s) \frac{\partial U(s)}{\partial \dot{k}^i(s)} \right]_{s=t} = - \rho \int_t^\infty e^{-\rho s} U(s) \, ds, \qquad (5.11a)$$

or

$$e^{-\rho t} \left[U(t) - \dot{k}^i(t) \frac{\partial U(t)}{\partial \dot{k}^i(t)} \right] = \rho \int_t^\infty e^{-\rho s} U(s) \, ds, \qquad (5.11b)$$

Multiplying both sides of (5.11b) by $e^{\rho t}$, we get the conservation law

$$U(t) - \dot{k}^i(t) \frac{\partial U(t)}{\partial \dot{k}^i(t)} = \rho \int_t^\infty e^{-\rho (s-t)} U(s) \, ds. \qquad (5.12a)$$

The left-hand side of the above is nothing but the utility measure of 'income,' while the right-hand side is the utility measure of 'wealth.' Hence, we have

$$\text{income } (t) = \rho \text{ wealth } (t), \qquad 0 \le t \le \infty. \qquad (5.12b)$$

For $\rho > 0$, we have the income-wealth conservation law [see also Samuelson (1982), Sato (1982), Weitzman (1976) and Kemp and Long (1982)].

$$\text{income } (t) \, / \, \text{wealth } (t) = \rho = \text{constant}. \qquad (5.12c)$$

We can give an alternative interpretation to (5.12b). Since $\int_t^\infty e^{-\rho(s-t)} U(s) \, ds$ is defined as wealth, the integral in the null term in equation (5.9b) may be called 'discounted wealth,' for we have

$$\Phi \Big|_\infty^t = \rho \int_t^\infty e^{-\rho s} U(s) = \rho \, e^{-\rho t} \text{ wealth } (t). \qquad (5.13a)$$

This, then, becomes

$$\Phi(t) - \Phi(\infty) = \rho \left[\int_t^0 e^{-\rho s} U(s)\, ds - \int_\infty^0 e^{-\rho s} U(s)\, ds \right]$$

(5.13b)

$$= \rho \left[\int_0^\infty e^{-\rho s} U(s)\, ds - \int_0^t e^{-\rho s} U(s)\, ds \right]$$

$$\Phi \Big|_\infty^\tau = \rho\, [S^* - S(t)].$$

(5.13c)

Here S^* = discounted total (maximum) stock of consumption measured in terms of utility and $S(t)$ = discounted stock of consumption also measured in terms of utility up to time t. For short, we shall call S, the stock of consumption. Using (5.9b), (5.11b) and (5.13b), we can express the coservation law as

$$e^{-\rho t} \left[U(t) - \dot{k}^i\, \frac{\partial U(t)}{\partial \dot{k}^i(t)} \right] + \rho \int_0^t e^{-\rho s} U(s)\, ds = \rho \int_0^\infty e^{-\rho s} U(s)\, ds.$$

(5.14a)

or

discounted income(t) + ρ × discounted stock of consumption
= ρ × maximum discounted stock of consumption
= constant.

(5.14b)

discounted income + discounted stock income
= modified income = constant.

(5.14c)

This is the alternative formulation of the income-wealth conservation law in terms of the modified income [see Sato (1981)].

6. SPECIAL CASES

It may be of interest to study special cases of the one capital good model of Ramsey (1928) and of Liviatan and Samuelson (1969). First, the Ramsey model corresponds to the case where $i = 1$ and U takes the form

$$U = U[k, \dot{k}] = U[c(k, \dot{k})] = U[f(k(t)) - \dot{k}(t)], \tag{6.1}$$

where c satisfies

$$U'(c) > 0, \qquad U''(c) < 0 \qquad \text{for} \qquad 0 < c < \infty, \tag{6.1a}$$

$$U'(0) = \infty, \tag{6.1b}$$

and f satisfies

$$f(k) > 0, \, f'(k) > 0, \, f''(k) < 0 \text{ for } 0 < k < \bar{k} < \infty, \tag{6.2a}$$

$$f(k) < f(\bar{k}) = \bar{c} \quad \text{for} \quad 0 < k < \bar{k}, \tag{6.2b}$$

$$f(k) \leq f(\bar{k}) \quad \text{for} \quad \bar{k} \leq k, \tag{6.2c}$$

where $\bar{k}$ = golden rule capital and $\bar{c}$ = bliss consumption. Then the income-wealth conservation law (5.12b) reduces to

$$U(t) + \dot{k}(t)\, U'(t) = \rho \int_t^\infty e^{-\rho(s-t)}\, U(s)\, ds. \tag{6.3a}$$

$$\begin{array}{ccc} \text{utility measure} & + & \text{utility measure} \\ \text{of consumption} & & \text{of investment} \end{array} = \begin{array}{c} \text{utility measure} \\ \text{of income} \end{array}$$

(6.3b)

$$= \rho \times \begin{array}{c} \text{utility} \\ \text{measure} \\ \text{of wealth} \end{array}$$

This special case is extensively discussed by Samuelson (1982) and Sato (1982).

Secondly, the Liviatan-Samuelson model corresponds to the case where $i = 1$ and U takes the general form

$$U = U[k, \dot{k}] = U(c) = U[c(k(t), \dot{k}(t))], \tag{6.4a}$$

$$\partial c / \partial k < 0, \quad \partial^2 c / \partial k^2 < 0, \tag{6.4b}$$

$$\max c(0, k) \quad \text{at} \quad k = \bar{k}, \tag{6.4c}$$

$$\partial c(0, k) / \partial k \gtrless 0 \quad \text{as} \quad k \lessgtr \bar{k}. \tag{6.4d}$$

Here the conservation law (5.12b) holds for $i = 1$ automatically.

7. GENERALIZED INCOME: WEALTH CONSERVATION LAWS

We now consider the most general case of utility maximization where the society's purpose is to maximize

$$\int_0^\infty W[t, k(t), \dot{k}(t)] \, dt, \qquad k(t) = (k^1(t), \ldots, k^n(t)), \tag{7.1}$$

subject to the appropriate initial conditions. More specifically W may be expressed as

$$W[t, k(t), \dot{k}(t)] = e^{-\rho(t)} U[t, k(t), \dot{k}(t)], \tag{7.2}$$

and W satisfies all the requirements of a social welfare function. In this general case the society may have: (1) variable discount rate $\rho(t)$, and (2) taste and technical change $\partial U / \partial t \neq 0$.

Noether's invariance condition for this case is now written as

$$\frac{\partial W}{\partial t} \tau + \frac{\partial W}{\partial k^i} \xi^i + \frac{\partial W}{\partial k^i} \left(\frac{d\xi^i}{dt} - \dot{k}^i \frac{d\tau}{dt} \right) + W \frac{d\tau}{dt} = \frac{d\Phi}{dt},$$

$$\Phi = \Phi(t, k). \tag{7.3}$$

Then for the optimal path, we have the conservation law expressed in terms of the time derivative,

$$\frac{d}{dt} \left[\left(W - \dot{k}^i \frac{\partial W}{\partial \dot{k}^i} \right) \tau + \frac{\partial W}{\partial \dot{k}^i} \xi^i - \Phi \right] = 0. \tag{7.4}$$

Using (7.3) and (7.4) we obtain the null term as

$$\frac{d\Phi}{dt} = \frac{d}{dt} \left[\left(W - \dot{k}^i \frac{\partial W}{\partial \dot{k}^i} \right) \tau + \frac{\partial W}{\partial \dot{k}^i} \xi^i \right]$$

$$\tag{7.5}$$

$$= \frac{\partial W}{\partial t} \tau + \frac{\partial W}{\partial k^i} \xi^i + \frac{\partial W}{\partial k^i} \left(\frac{d\xi^i}{dt} - \dot{k}^i \frac{d\tau}{dt} \right) + W \frac{dt}{dt}.$$

This is the fundamental equation of the conservation law, which contains many special cases depending upon the forms of τ and ξ^i. Let

$\xi^i = 0$ for $i = 1, \ldots, n$. Then integrating (7.5) we have the *generalized income -wealth conservation law*

$$\frac{d}{dt}\left[\left(W - \dot{k}^i \frac{\partial W}{\partial \dot{k}^i}\right)\tau\right] = \frac{\partial W}{\partial t}\tau + \left(W - \dot{k}^i \frac{\partial L}{\partial \dot{k}^i}\right)\frac{d\tau}{dt}, \qquad (7.6a)$$

$$\left[\left(W - \dot{k}^i \frac{\partial W}{\partial \dot{k}^i}\right)\tau\right]_{t=T}^{b} = \int_T^b \left[\frac{\partial W}{\partial t}\tau + \left(W - \dot{k}^i \frac{\partial W}{\partial \dot{k}^i}\right)\frac{d\tau}{dt}\right]dt, \quad (7.6b)$$

$$\left[\left(W - \dot{k}^i \frac{\partial W}{\partial \dot{k}^i}\right)\tau\right]_{t=T} = -\int_T^{\infty} \left[\frac{\partial W}{\partial t}\tau + \left(W - \dot{k}^i \frac{\partial W}{\partial \dot{k}^i}\right)\frac{d\tau}{dt}\right]dt, \quad (7.6c)$$

where we assumed, by the transversality condition, that

$$\left(W - \dot{k}^i \frac{\partial W}{\partial \dot{k}^i}\right)\tau \to 0, \qquad \frac{\partial W}{\partial t}\tau + \left(W - \dot{k}^i \frac{\partial W}{\partial \dot{k}^i}\right)\frac{d\tau}{dt} \to 0$$

$$\text{as } t \to \infty.$$

The left-hand side of (7.6c) represents *generalized income* with the time transformation τ, while the right-hand side of (7.6c) represents *generalized wealth*. In particular, when

$$\tau = 1, \qquad (7.7a)$$

we have

$$\left(W - \dot{k}^i \frac{\partial W}{\partial \dot{k}^i}\right) = \text{utility measure of generalized income,} \quad (7.7b)$$

and

$$-\int_t^{\infty} \frac{\partial W}{\partial s}\,ds = \text{utility measure of generalized wealth.} \quad (7.7c)$$

The economic interpretation of (7.7c) as generalized wealth will become more apparent when we consider a special case of W in (7.2) as

$$W = e^{-\rho(t)} U[k(t), \dot{k}(t)]. \tag{7.8a}$$

In this case $-\int_t^\infty (\partial W / \partial s)\, ds$ is equal to

$$-\int_t^\infty \frac{\partial W}{\partial s}\, ds = \int_t^\infty \rho'(s)\, e^{-\rho(s)}\, U[k(s), \dot{k}(s)]\, ds. \tag{7.8b}$$

The standard income-wealth conservation law (5.12a) can be derived when $\rho(t)$ takes a special form,

$$\rho(t) = \rho t, \tag{7.8c}$$

and

$$-\int_t^\infty \frac{\partial W}{\partial s}\, ds = \rho \int_t^\infty e^{-\rho s}\, U[k(s), \dot{k}(s)]\, ds. \tag{7.8d}$$

Another special case of interest occurs when τ and W take the special forms

$$\tau = 1 / \rho'(t), \tag{7.9a}$$

and

$$W = e^{-\rho(t)} U[k(t), \dot{k}(t)]. \tag{7.9b}$$

After some somewhat complicated calculations, we obtain the generalized income-wealth conservation law as

$$\left[U - \dot{k}^{\,i}\, \frac{\partial U}{\partial \dot{k}^{\,i}} \right] = \rho_t \int_t^\infty \exp\left(-\int_t^s \rho_p\, dp \right) \left[U - \dot{k}^{\,i}\, \frac{\partial U}{\partial \dot{k}^{\,i}}\, \frac{d}{ds}\left(\frac{1}{\rho_s} \right) \right] ds, \quad (7.9c)$$

$$\text{income} \;=\; \rho_t \times \text{generalized wealth,} \qquad\qquad (7.9)$$

where

$$\rho_t \;=\; d\rho\,/\,dt \;=\; \rho'(t).$$

The generalized wealth expression now includes capital gains and losses, depending upon whether $-k^{\,i}(\partial U\,/\,\partial k^{\,i})$ $(d\,/\,ds)$ $(1\,/\,\rho_s)$ is positive or negative. This term depends on the supply price of investment, $-\partial U\,/\,\partial k$, and the *variable* discount rate, $\rho_s(s)$. Needless to say that when $d\,/\,ds\,(\rho_s) \equiv 0$, i.e., the constant discount rate, (7.9c) reduces to the standard income-wealth conservation law (5.12a). Hence, when $\rho = $ constant, the time transformations $\tau = 1$ and $\tau = 1\,/\,\rho$ give the identical results.

Next we consider the effect of taste and technical change in the utility and production functions. The generalized income-wealth conservation law contains the case of the welfare function

$$W = e^{-\rho t}\, U[t,\, k(t),\, \dot{k}(t)], \qquad\qquad (7.10a)$$

where U is subjected to taste and technical change so that

$$\partial U\,/\,\partial t \neq 0. \qquad\qquad (7.10b)$$

The null term in this case is simply equal to

$$\frac{d\Phi}{dt} = e^{-\rho t}\left[-\rho U + \frac{\partial U}{\partial t} \right]. \qquad\qquad (7.10c)$$

Hence, the conservation law when $\tau = 1$ and $\xi^i = 0$ becomes

$$\left[U - \dot{k}^i \frac{\partial U}{\partial \dot{k}^i} \right] + \int_t^\infty e^{-\rho(s-t)} \frac{\partial U}{\partial s} ds = \rho \int_t^\infty e^{-\rho(s-t)} U(s) ds. \quad (7.10d)$$

income + current worth of taste (technical) change
$= \rho \times$ wealth. $\hspace{4cm}$ (7.10e)

modified income $= \rho \times$ wealth. $\hspace{3cm}$ (7.10f)

The value of wealth can be expressed in the same form as in the case when $\partial U / \partial t = 0$, but income must be modified by the amount of taste (technical) change. Note that in deriving (7.10d), we used the transversality condition

$$e^{-\rho t} \left[U - \dot{k}^i \frac{\partial U}{\partial \dot{k}^i} \right] \to 0 \quad \text{as} \quad t \to \infty.$$

8. INCOME-CAPITAL (WEALTH) CONSERVATION LAW IN THE VON NEUMANN MODEL

As this problem has already been studied extensively by Samuelson (1970a,b) and Sato (1981), we shall briefly present the results. The von Neumann problem is to maximize

$$\int_0^T \dot{K}_1 \, dt \text{ subject to } F[K_1, ..., K_n, \dot{K}_1, ..., \dot{K}_n] = 0, \quad (8.1a)$$

and subject to the appropriate initial conditions. Here K_i are the ith type of capital, $\dot{K}_i$ are the net capital formation of the ith type of capital ($i = 1, ... , n$) and F is a smooth, neoclassical, concave first-degree-homogeneous transformation function.

The Lagrangean associated with (8.1a) is

$$L[t, \lambda, K_1, ..., K_n, \dot{K}_1, ..., \dot{K}_n, \dot{\lambda}]$$

$$= \dot{K}_1 + \lambda F[K_1, ..., K_n, \dot{K}_1, ..., \dot{K}_n] \tag{8.1b}$$

$$= \dot{K}_1 + \lambda F[K, \dot{K}], \tag{8.1c}$$

where

$$K = (K_1, ..., K_n).$$

The transformations under consideration are

$$\bar{t} = t + \varepsilon\tau(t, \lambda, K),$$

$$\bar{K}_i = K_i + \varepsilon\xi^i(t, \lambda, K), \quad i = 1, ..., n. \tag{8.2}$$

$$\bar{\lambda} = \lambda + \varepsilon\omega(t, \lambda, K),$$

The Noether invariance principle requires that

$$\frac{d\Phi}{dt} = F\omega + \frac{d\xi^1}{dt} + \lambda\left(\frac{\partial F}{\partial K_i}\xi^i + \frac{\partial F}{\partial \dot{K}_i}\frac{d\xi^i}{dt} + K_i\frac{\partial F}{\partial K_i}\frac{d\tau}{dt}\right). \tag{8.3}$$

The conservation laws are derived from

$$\Omega = \lambda\left(F - \dot{K}_i\frac{\partial F}{\partial \dot{K}_i}\right)\tau + \xi^I + \lambda\frac{\partial F}{\partial \dot{K}_i}\xi^i - \Phi = \text{constant.} \tag{8.4}$$

Let $\tau = 1$, $\omega = 0$, $\xi^i = 0$, and $d\Phi / dt = 0$, i.e., $\Phi = \text{constant} = c$, we then have

$$\Omega = \lambda \left(F - \dot{K}_i \frac{\partial F}{\partial \dot{K}_i} \right) - c = \lambda \dot{K}_i \frac{\partial F}{\partial \dot{K}_i} - c = \text{constant}, \quad (8.5a)$$

where

$$F = K_i \frac{\partial F}{\partial K_i} + \dot{K}_i \frac{\partial F}{\partial \dot{K}_i},$$

by homogeneity. Letting $Y = K_i (\partial F / \partial K_i) = \text{national income}$, the above can be interpreted as

$$\Omega_1 = \lambda Y = \text{constant}. \quad (8.5b)$$

Now let us assume that $\tau = 0$ and $\xi^i = \alpha K_i$ and $\omega = -\alpha \lambda$. Then we have

$$\frac{d\Phi}{dt} = -\alpha \lambda F + \alpha \dot{K}_1 + \alpha \lambda \left(K_i \frac{\partial F}{\partial K_i} + \dot{K}_i \frac{\partial F}{\partial \dot{K}_i} \right) = \alpha \dot{K}_1. \quad (8.6a)$$

Hence, we have

$$\Phi = \alpha K_1 + c, \qquad c = \text{constant}, \quad (8.6b)$$

which gives

$$\Omega_2 = \alpha K_1 + \alpha \lambda K_i \frac{\partial F}{\partial \dot{K}_i} - (\alpha K_1 + c)$$

$$\quad (8.6c)$$

$$= \alpha \lambda K_i \frac{\partial F}{\partial \dot{K}_i} - c = \text{constant},$$

or

$$\Omega_2 = -\lambda K_i \frac{\partial F}{\partial \dot{K}_i} = \lambda W = \text{constant},$$ (8.6d)

where

$$W = -K_i \frac{\partial F}{\partial \dot{K}_i} = K_i P^i = \text{national wealth.}$$

Using (8.5b) and (8.6d), we have the Samuelson conservation law

$$\frac{\Omega_1}{\Omega_2} = \frac{Y}{W} = \frac{\text{national income}}{\text{national wealth}} = \text{constant.}$$ (8.7)

The conservation law (8.7) is a global law independent of the form of technology. But if we allow for special forms, there may exist different types of conservation law in addition to (8.7). In my earlier work (Sato, 1981, p. 289), I posed an open question regarding the existence of any new conservation law for a special separable technology

$$F = Y(K) - I(\dot{K}) = 0,$$ (8.8a)

$$F = K_1^{a_1} K_2^{a_2} \ldots K_n^{a_n} - (\dot{K}_1^2 + \dot{K}_2^2 + \ldots \dot{K}_n^2)^{1/2} = 0$$ (8.8b)

with

$$\sum_{i=1}^{n} a_i = 1$$

This form of technology has been used by many economists including Samuelson (1970b), Caton and Shell (1971). Kataoka

(1983) has now shown that (8.7) is the only conservation law for this special case. Thus my question has been answered in the negative.

Let us briefly present Kataoka's arguments. To simplify the calculations, we set $i = 1, 2$. Then the invariance principle requires [Sato (1981, p. 289, eq. (109))] that

$$\frac{\partial \xi^1}{\partial K_1} = \frac{\partial \xi^2}{\partial K_2}, \quad \frac{\partial \xi^1}{\partial K_2} + \frac{\partial \xi^2}{\partial K_1} = 0 \tag{8.9a}$$

$$K_1 K_2 \left(\alpha - \frac{\partial \xi^1}{\partial K_1} \right) + aK_2 \, \xi^1 + (1-a) \, K_1 \, \xi^2 = 0, \tag{8.9b}$$

where $a = a_I$ in (8.8b) and α is the coefficient of the time transformation $\tau = \alpha t + \beta$. Rewriting the first part of (8.9a) as $\partial \xi^1 / \partial K_1 = 1/2 (\partial \xi^1 / \partial K_1 + \partial \xi^2 / \partial K_2)$ and substituting it into (8.9b), we have

$$K_1 K_2 \left[\frac{1}{2}\alpha + \frac{1}{2} \alpha - \frac{1}{2} \left(\frac{\partial \xi^1}{\partial K_1} + \frac{\partial \xi^2}{\partial K_2} \right) \right] + aK_2 \, \xi^1 + (1-a) \, K_1 \, \xi^2 = 0. \tag{8.9c}$$

This yields

$$\frac{\alpha}{2} - \frac{1}{2} \frac{\partial \xi^1}{\partial K_1} + \frac{a\xi^1}{K_1} = -\frac{\alpha}{2} + \frac{1}{2} \frac{\partial \xi^2}{\partial K_2} - (1-a)\frac{\xi^2}{K_2}. \tag{8.9d}$$

Assuming $\xi^1 = \xi^1 (K_1)$ and $\xi^2 = \xi^2 (K_2)$, (8.9d) yields two ordinary differential equations

$$\frac{\alpha}{2} - \frac{1}{2} \frac{\partial \xi^1}{\partial K_1} + a \frac{\xi^1}{K_1} = B = \text{constant}, \tag{8.9e}$$

$$-\frac{\alpha}{2} + \frac{1}{2} \frac{\partial \xi^2}{\partial K_2} - (1-a) \frac{\xi^2}{K_2} = B = \text{constant}.$$

Among the solutions of (8.9e), the only relevant ones which satisfy (8.9a) and (8.9b) are

$$\alpha = 0, \qquad \xi^1 = \gamma K_1, \qquad \xi^2 = \gamma K_2, \qquad (8.9f)$$

which proves that (8.7) is the only conservation law even for this special technology. Kataoka also shows that even if we specify $\xi^i (K_1, K_2)$ the final solution yields the same result.

9. A REMARK OF ECONOMETRIC APPLICATIONS

The invariance principle and conservation laws may serve as useful devices for the purpose of empirically testing the validity of optimal growth models. Although much work has been done in recent years regarding the problem of optimal growth, little effort has been made to determine whether or not optimal growth models have any value as empirically explanatory tools—except the works of Lenard (1972), Sakakibara (1970) and a few others.

Both Lenard and Sakakibara tested the empirical validity of their optimal growth models by comparing the simulated optimal time paths for the capital-labor and saving ratios with the actual time paths of these ratios [e.g., see Lenard (1972, ch.4)]. They used the sums of the deviation squared as the criterion for the test of their models.

The conservation laws discussed earlier can be used to indirectly check the validity of such models as Lenard and Sakakibara. For instance, by using reasonable utility functions, as in Lenard and Sakakibara, one can empirically construct the time series for the income-wealth ratio. If the series are relatively stable, this would be an indication that the models work and that the actual paths follow closely the optimal paths presented by the models. In the income-wealth equation,

$$\text{income} \,/\, \text{wealth} = f(t) = a + bt,$$

one may propose to test the hypothesis that the slope coefficient b is not statistically different from zero.

Of course, the actual paths will be affected by both technical and taste changes. Consequently, the test equation may be much more complicated than the one suggested above. Nevertheless, it should be apparent that knowledge of conservation laws could by of some assistance in the econometric testing of optimal growth models. This is an important and interesting area of research to which I plan to turn my attention in the future.

10. SUMMARY

In this paper by using the Noether invariance theorem we have uncovered several 'hidden' conservation laws in the model with heterogeneous capital goods. Our first conservation law is a pseudo-net-productivity relation which implies that the rate of change in national income is equal to the discount rate multiplied by the utility-value-of-investment. The second conservation law is an integral version of the first law—the constancy of income-wealth ratio. It is shown that the second law can be interpreted as the law of variance of 'modified income.'

If we intoduce taste change and/or technical change it can be proved that there exist several generalized income-wealth conservation laws. Income and wealth must now include the effects of taste and technical change. Also under a *variable* discount rate, income-wealth conservation laws contain terms related to capital gains.

The analysis is extended to the von Neumann model of capital accumulation. Here again we confirm the existence of the income-capital conservation law. Even if technology is limited to a specialized type, there exist no additional invariances, other than those discovered by Samuelson (1970) and Sato (1981).

From an empirical perspective, knowledge of conservation laws can assist in the econometric testing of optimal growth models. In addition to the capital-labor, and saving ratios, we can now examine the behavior of the appropriate income-wealth ratios. This additional dimension should be of some value in empirical research on optimal growth models.

APPENDIX

Consider the variational integral

$$J(x) = \int_a^b L(t, x(t), \dot{x}(t)) \, dt, \qquad \begin{aligned} x(t) &= (x^1(t), \ldots, x^n(t)). \\ \dot{x}(t) &= (\dot{x}^1(t), \ldots, \dot{x}^n(t)). \end{aligned} \qquad \text{(A.1)}$$

Also consider r-parameter transformations

$$T: \quad \begin{aligned} \bar{t} &= \phi(t, x; \varepsilon), \quad \varepsilon = (\varepsilon^1, \ldots, \varepsilon^n), \\ \bar{x}^i &= \psi^i(t, x; \varepsilon), \quad i = 1, \ldots, n. \end{aligned} \qquad \text{(A.2)}$$

where

$$\phi(t, x; 0) = t,$$
$$\psi^i(t, x; 0) = x^i. \qquad \text{(A.3)}$$

In addition, it is assumed that T does not change the end points (a, α) and (b, β) where

$$\phi(a, x(a); \varepsilon) = a, \qquad \phi(b, x(b); \varepsilon) = b,$$
$$\psi^i(a, x(a); \varepsilon) = \alpha^i, \qquad \psi^i(b, x(b); \varepsilon) = \beta^i. \qquad \text{(A.4)}$$

Let X_s be the infinitesimal transformations for $s = 1, ..., r$, then we write X_s as

$$X_s = \tau_s(t, x) \frac{\partial}{\partial t} + \xi_s^i(t, x) \frac{\partial}{\partial x^i} + \eta_s^i(t, x, \dot{x}) \frac{\partial}{\partial \dot{x}^i}, \qquad (A.5)$$

where

$$\tau_s(t, x) = \frac{\partial \phi}{\partial \varepsilon^s}(t, x; 0), \qquad \xi_s^i(t, x) = \frac{\partial \psi^i}{\partial \varepsilon^s}(t, x; 0),$$

$$(A.6)$$

$$\eta_s^i(t, x, x) = \frac{d\xi_s^i}{dt} - x^i \frac{d\tau_s}{dt},$$

and (A.4) obeys

$$\tau_s(a, x(a)) = \tau_s(b, x(b)) = 0,$$

$$(A.7)$$

$$\xi_s^i(a, x(a)) = \xi_s^i(b, x(b)) = 0.$$

If (A.1) is invariant under (A.2), we have

$$\int_a^b X_s(L\, dt) = 0. \qquad (A.8)$$

Lemma. *For any $\tau_s(t, x)$ and $\xi_s^i(t, x)$, we have*

$$N_s = (\xi_s^i - \dot{x}^i \tau_s) E_i + \frac{d\Omega_s}{dt}, \qquad (A.9)$$

where

$$N_s = \frac{\partial L}{\partial t}\, \tau_s + \frac{\partial L}{\partial x^i}\, \xi_s^i + \frac{\partial L}{\partial \dot{x}^i}\left(\frac{d\xi_s^i}{dt} - \dot{x}^i\, \frac{d\tau_s}{dt}\right) + L\, \frac{d\tau_s}{dt}\ ,$$

$$E_i = \text{Euler-Lagrange equation} = \frac{\partial L}{\partial x^i} - \frac{d}{dt}\left(\frac{\partial L}{\partial \dot{x}^i}\right),$$

$$\frac{d\Omega}{dt} = \frac{d}{dt}\left[\left(L - \dot{x}^i\, \frac{\partial L}{\partial \dot{x}^i}\right)\tau_s + \frac{\partial L}{\partial \dot{x}^i}\, \xi_s^i\right].$$

Proof. By differentiating Ω_s with respect to t we have

$$\frac{d\Omega_s}{dt} = \frac{d}{dt}\left[\left(L - \dot{x}^i\, \frac{\partial L}{\partial \dot{x}^i}\right)\tau_s + \frac{\partial L}{\partial \dot{x}^i}\, \xi_s^i\right]$$

$$= \frac{\partial L}{\partial t}\, \tau_s + \frac{\partial L}{\partial \dot{x}^i}\left(\frac{d\xi_s^i}{dt} - \dot{x}^i\, \frac{d\tau_s}{dt}\right) + L\, \frac{d\tau_s}{dt}$$

$$+ x^i\, \tau_s\left[\frac{\partial L}{\partial x^i} - \frac{d}{dt}\left(\frac{\partial L}{\partial \dot{x}^i}\right)\right] + \frac{d}{dt}\left(\frac{\partial L}{\partial \dot{x}^i}\right)\xi_s^i$$

$$= N_s - (\xi_s^i - \dot{x}^i\, \tau_s)\, E_i\,.\qquad\qquad\text{Q.E.D.}$$

When (A.1) is optimized, E_i vanishes and (A.9) reduces to

$$d\Omega_s\, /\, dt = N_s\,.\qquad\qquad\text{(A.10)}$$

There are two cases: (1) when $N_s = 0$ and (2) $N_s \neq 0$. The conservation law when $N_s = 0$ is

$$d\Omega_s / dt = 0 \quad \text{or} \quad \Omega_s = \text{constant.} \tag{A.11a}$$

When $N_s \neq 0$ and $d\Phi_s / dt = N_s$, we have

$$d(\Omega_s - \Phi_s) / dt = 0 \quad \text{or} \quad \Omega_s - \Phi_s = \text{constant.} \tag{A.11b}$$

This is Noether's invariance up to divergence.

NOTES

1. Her paper has recently been translated into English by Tavel, who also supplies a brief motivation and historical sketch (see Noether, 1918).
2. See Klein (1918), Lie (1891), Gelfand and Fomin (1963), Rund (1966), Logan (1977), Nôno (1968), Nôno and Mimura (1975, 1976, 1977, 1978), Sagan (1969), Lovelock and Rund (1975), Whittaker (1937), and Moser (1979).
3. A comprehensive treatment of Noether's theorem and invariance identities is given in Sato (1981, pp. 236–251). Those who are familiar with this aspect of the mathematics can skip this section.
4. We adopt the so-called Einstein summation convention (Young, 1978, pp. 333–334). When a lower case Latin index such as $j, h, k, \ldots$ appears twice in a term, summation over that index is implied, the range of summation being $1, \ldots, n$. For example, $a^i x^i$ means $\sum_{i=1}^{n} a^i x^i$.
5. In many cases the transformations (4.2) may be Lie groups and, hence, (4.5) may be the infinitesimal transformations of a group. However, to study Noether's invariance principle the group property is *not* necessary. We simply assume the existence of the infinitesimal transformations.

REFERENCES

Bessel-Hagen, E. (1921). Über die Erhaltungssätze der Elektrodynamik. *Math. Ann., 84,* 258–276.

Caton, C. and Shell, K. (1971). An exercise in the theory of heterogeneous capital accumulation. *Review of Economic Studies, 32,* 233–240.

Gelfand, I. M and Fomin, S. V. (1963). *Calculus of variations.* Englewood Cliffs, NJ: Prentice-Hall, translated from the Russian by R. A. Silverman.

Kataoka, H. (1983). On the local conservation laws in the von Neumann model. In R. Sato and M. J. Beckmann (eds.), *Technology, organization and economic structure: Lecture notes in economics and mathematical systems.* Vol. 210, 253–260.

Klein, F. (1918). Über die Differentialgesetze für die Erhaltung von Impuls und Energie in der Einsteinschen Gravitationstheorie, Nachr. Akad. Wiss. Göttingen, Math-Phys. Kl. II, 171–189.

Kemp, M. C. and Long, N. V. (1982). On the evaluation of social income in a dynamic economy. In G. R. Feiwel (ed.), *Samuelson and neoclassical economics.* Boston: Kluwer-Nijhoff.

Lenard. T. M. (1972). *Aggregate policy controls and optimal growth: Theory and applications to the U.S. economy.* Providence, RI: Brown University (PhD dissertation).

Lie, S. (1891). Vorlesungen über Differentialgleichungen, mit bekannten infinitesimalen Transformationen. Edited by G. Scheffers, Leipzig: Teubner. Reprinted (1967) New York: Chelsea.

Liviatan, N. and Samuelson, P. A. (1969). Notes on turnpikes: Stable and unstable. *Journal of Economic Theory, 1,* 454–475.

Logan, J. D. (1977). Invariant variational principles. *Mathematics in science and engineering.* Vol. 138. New York: Academic Press.

Lovelock D. and Rund, H. (1975). *Tensors, differential forms and variational principles.* New York: Wiley.

Moser, J. (1979). Hidden symmetries in dynamical systems. *American Scientists, 67,* 689–695.

Noether, E. (1918). Invariante Variationsprobleme, Nachr. Akad. Wiss. Göttingen, Math-Phys. Kl. II, 235–257. Translated by M. A. Tavel (1971). Invariant variation problems. *Transport Theory and Statistical Physics, 1,* 186–207.

Nôno, T. (1968). On the symmetry groups of simple materials: applications of the theory of Lie groups. *J. Math. Anal. Appl., 24,* 110–135.

Nôno, T. and Mimura, F. (1975/1976/1977/1978). Dynamic symmetries I/III/IV/V. *Bulletin of Fukuoka University of Education, 25/26/27/28.*

Ramsey, F. (1928). A mathematical theory of saving. *Economic Journal, 38,* 543–559.

Rund, H. (1966). *The Hamilton–Jacobi theory in the calculus of variations.* Princeton, NJ: Van Nostrand–Reinhold.

Sagan, H. (1969). *Introduction to the calculus of variations.* New York: McGraw-Hill.

Sakakibara, E. (1970). Dynamic optimization and economic policy. *American Economic Review, 60.*

Samuelson, P. A. (1970a). Law of conservation of the capital-output ratio: Proceedings of the National Academy of Sciences. *Applied Mathematical Science, 67,* 1477–1479.

Samuelson, P. A. (1970b). To conservation laws in theoreticl economics. Cambridge, MA: M.I.T. Department of Economics mimeo.

Samuelson, P. A. (1972). The general saddlepoint property of optimal-control motions. *Journal of Economic Theory, 5,* 102–120.

Samuelson, P.A. (1976). Speeding up of time with age in recognition of life as fleeting. In A.M. Tang *et al.* (eds.) *Evolution, welfare, and time in economics: Essays in honor of Nicholas Georgescu-Roegen.* Lexington, MA: Lexington/ Heath Books.

Samuelson, P.A. (1982). Variations on capital-output conservation laws. Cambrige, MA: M.I.T. mimeo.

Sato, R. (1981). *Theory of technical change and economic invariance: Application of Lie groups.* New York: Academic Press.

Sato, R. (1982). Invariant principle and capital-output conservation laws. Providence, RI: Brown University working paper No. 82–8.

Sato, R., Nôno, T. and Mimura, F. (1983). Hidden symmetries: Lie groups and economic conservation laws, Essay in honor of Martin Beckmann.

Weitzman, M. L. (1976). On the welfare significance of national product in a dynamic economy. *Quarterly Journal of Economics, 90,* 156–162.

Whittaker, E. T. (1937). A treatise on the analytical dynamics of particles and rigid bodies. Cambridge: Cambridge University Press, 4th edition. Reprinted (1944), New York: Dover.

CHAPTER 5

Conservation Laws Derived *via* the Application of Helmholtz Conditions

Fumitake Mimura
Takayuki Nôno *

I. INTRODUCTION

The inverse problem of Lagrangean dynamics in the variational principle is to establish necessary and sufficient conditions for a given system of n differential equations

$$G_i\,(\ddot{x}, \dot{x}, x, t) = G_i\,(\ddot{x}_1, ..., \ddot{x}_n; \dot{x}_1, ..., \dot{x}_n; x_1, ..., x_n; t) = 0 \quad (1)$$

to be identified with the Euler-Lagrange equations for some Lagrangean $L(\dot{x}, x, t)$:

$$[L]_i = \frac{d}{dt}\left(\frac{\partial L}{\partial \dot{x}_i}\right) - \frac{\partial L}{\partial x_i} = 0$$

where $x_i = x_i\,(t)$, $\dot{x}_i = dx_i\,/\,dt$ and $\ddot{x}_i = d^2 x_i\,/\,dt^2$ ($i = 1, ..., n$). The conditions, usually referred to as the Helmholtz conditions due to the initial investigation of Helmholtz (1887), are as follows:

* Department of Mathematics, Faculty of Engineering, Kyushu Institute of Technology, and Department of Mathematics, Fukuyama University, respectively. F. Mimura would like to acknowledge a Grant-in-Aid for Scientific Research (No. 63540059) from the Ministry of Education.

$$\frac{\partial G_i}{\partial \ddot{x}_j} = \frac{\partial G_j}{\partial \ddot{x}_i}, \tag{2a}$$

$$\frac{\partial G_i}{\partial \dot{x}_j} + \frac{\partial G_j}{\partial \dot{x}_i} = \frac{d}{dt}\left(\frac{\partial G_i}{\partial \ddot{x}_j} + \frac{\partial G_j}{\partial \ddot{x}_i} \right), \tag{2b}$$

$$\frac{\partial G_i}{\partial x_j} - \frac{\partial G_j}{\partial x_i} = \frac{1}{2}\frac{d}{dt}\left(\frac{\partial G_i}{\partial x_j^{\cdot}} - \frac{\partial G_j}{\partial x_i^{\cdot}} \right). \tag{2c}$$

It follows immediately from (2a) and (2b) that G_i must be linear in the $\ddot{x}_i$. Hence, the equations (1) may be assumed without loss of generality to be of the form:

$$G_i\,(\ddot{x},\,\dot{x},\,x,\,t) = G_{ij}\,(\dot{x},\,x,\,t)\,\ddot{x}_j + g_i\,(\dot{x},\,x,\,t) = 0. \tag{3}$$

Helmholtz did not prove the sufficiency of his conditions. However, the sufficiency of the Helmholtz conditions was provided later by different authors: for the case $n = 1$, by Darboux (1894) and expanded to higher order derivatives by Hirsch (1898) and Boehm (1900); and for the general case $n > 1$, by Mayer (1896), Königsberger (1901), Hamel (1903), Kürschak (1905), Davis (1928), Douglas (1941), and Havas (1957). These authors essentially imposed on equations (3) the regularity condition $\det(G_{ij}) \neq 0$; which was removed by Engels (1975), who established the sufficiency proof of the Helmholtz conditions in terms of alternative forms. And subsequently Santilli (1977a, 1977b) carried the problem into the system of tensorial field equations (see also Santilli, 1978).

Independently of the inverse problem, from the equivalent Euler-Lagrangean equations $[L]_i = 0$ and $[N]_i = 0$ with regular Lagrangeans L and N, the following linear relations were derived:

$$[N]_i = C_i^j\,(\dot{x},\,x,\,t)\,[L]_j \tag{4}$$

in which the conservation laws (constants of the motion)

$$\text{tr}(C^j_i)^k = \text{const.}, \qquad \text{for integer } k > 0 \qquad (5)$$

were discovered: for the case $n = 1$, by Currie and Saletan (1966); and for the case $n > 1$, including the result of Lutzky [1979a, 1979b: det $(C^j_i) = \text{const.}$], by Hojman and Harleston (1981), which was revised in a more geometrical fashion by Henneaux (1981). Their results were reconstructed in the characteristic polynomial of the matrix (C^j_i) by Mimura and Nôno (1984), who pointed out that (4) implies (5) without the regularity conditions for the Lagrangians.

The existence of linear relations (4) can be conceived of as the inverse problem of a system of differential equations

$$C^j_i (x, \dot{x}, t) [L]_j = 0.$$

From this perspective, the Helmholtz conditions were applied to reconstruct the constants of the motion (5) by González-Gascón (1982) and by Farrias and Nergi (1983); which were generalized in field theory by Farrias and Teixeira (1983), while more general conserved identities in continuum dynamics were constructed by Mimura and Nôno (1986). (See also Mimura and Nôno, 1985.) Thus the Helmholtz conditions will play an important role in the derivation of conservation laws, including those that can not be obtained from the Noether (1918) symmetries (non-Noether conservation laws, Lutzky, 1979; Mimura and Nôno, 1985 and 1986.)

There were many different approaches to the inverse problem of Lagrangian dynamics. Nevertheless it is now being developed by several authors, for example, Henneaux (1982a, 1982b), Crampin (1981), Hojman and Urrutia (1981), Hojman (1984), Sarlet (1981, 1982), Sarlet and Cantrijn (1983), and others.

This chapter offers an illustrative usage of the Helmhotz conditions for deriving conservation laws (constants of the motion). The sufficiency proof of the Helmholtz conditions completed by Engels (1975) is reviewed with some modification. With the aid of the Helmholtz conditions, a new differential equation for the constants of the motion is constructed from a given Lagrangean in the case $n = 1$. The equation is used in neoclassical optimal growth models: simple model of the Ramsey (1928) type, the Liviatan-Samuelson (1969) model and the model suggested by Samuelson (1972). While some conservation laws in the models were obtained by Sato (1981) under Noether (1918) symmetries; more conservations laws, including non-Noether ones, are derived.

II. THE HELMHOLTZ CONDITIONS

We shall first reform the sufficiency proof of the Helmholtz conditions provided by Engels (1975). For convenience, differentiability is assumed to be of sufficiently high order and the summation convention is employed throughout. As stated in the introduction, the proof can be deduced without loss of generality for the restricted system of n differential equations of the form (3), in which the regularity condition $\det(G_{ij}) \neq 0$ is not assumed. Thus, the Helmholtz conditions are reduced to:

$$G_{ij} = G_{ji}, \tag{6a}$$

$$\frac{\partial G_{ik}}{\partial \dot{x}_j} = \frac{\partial G_{jk}}{\partial \dot{x}_i}, \tag{6b}$$

$$\frac{\partial g_i}{\partial \dot{x}_j} + \frac{\partial g_j}{\partial \dot{x}_i} = 2\left(\frac{\partial G_{ij}}{\partial x_k} \dot{x}_k + \frac{\partial G_{ij}}{\partial t} \right), \tag{6c}$$

$$\frac{\partial^2 g_i}{\partial \dot{x}_k \partial \dot{x}_j} - \frac{\partial^2 g_k}{\partial \dot{x}_i \partial \dot{x}_j} = 2\left(\frac{\partial G_{ij}}{\partial x_k} - \frac{\partial G_{kj}}{\partial x_i} \right), \tag{6d}$$

$$\frac{\partial g_i}{\partial x_j} - \frac{\partial g_j}{\partial x_i} = \frac{1}{2}\left(\frac{\partial^2 g_i}{\partial \dot{x}_j \partial x_k} - \frac{\partial^2 g_j}{\partial \dot{x}_i \partial x_k}\right)\dot{x}_k$$

$$+ \frac{1}{2}\left(\frac{\partial^2 g_i}{\partial \dot{x}_j \partial t} - \frac{\partial^2 g_i}{\partial \dot{x}_i \partial t}\right).$$

(6e)

Note that equation (6d) is derived as follows: differentiate (6c) by $\dot{x}_k$ and interchange the indices i and k; then the difference of the two resulting equations:

$$\frac{\partial^2 g_i}{\partial \dot{x}_k \partial \dot{x}_j} - \frac{\partial^2 g_k}{\partial \dot{x}_i \partial \dot{x}_j} = 2\left(\frac{\partial G_{ij}}{\partial x_k} - \frac{\partial G_{kj}}{\partial x_i}\right)$$

$$+ 2\left(\dot{x}_m \frac{\partial}{\partial x_m} + \frac{\partial}{\partial t}\right)\left(\frac{\partial G_{ij}}{\partial \dot{x}_k} - \frac{\partial G_{kj}}{\partial \dot{x}_i}\right)$$

reduces to (6d), since the last term vanishes by virtue of (6a).

The conditions (6a) and (6b) are necessary and sufficient for the existence of functions $\Phi_i(\dot{x}, x, t)$, such that

$$G_{ij} = \frac{\partial \Phi_i}{\partial \dot{x}_j} = \frac{\partial \Phi_j}{\partial \dot{x}_i},$$

which guarantees the existence of a function $\Phi(\dot{x}, x, t)$, such that

$$\Phi_i = \frac{\partial \Phi}{\partial \dot{x}_i}.$$

Hence the functions G_{ij} are of the forms

$$G_{ij} = \frac{\partial^2 \Phi}{\partial \dot{x}_i \partial \dot{x}_j},$$

(7)

and so the condition (6c) is translated into

$$\frac{\partial g_i}{\partial \dot{x}_j} + \frac{\partial g_j}{\partial \dot{x}_i} = 2 \left(\frac{\partial^3 \Phi}{\partial \dot{x}_i \partial \dot{x}_j \partial x_k} \dot{x}_k + \frac{\partial^3 \Phi}{\partial \dot{x}_i \partial \dot{x}_j \partial t} \right). \qquad (8)$$

Now define a new function $\phi_i \, (\dot{x}, x, t)$ by

$$\phi_i = g_i - \frac{\partial^2 \Phi}{\partial \dot{x}_i \partial x_k} \dot{x}_k - \frac{\partial^2 \Phi}{\partial \dot{x}_i \partial t} + \frac{\partial \Phi}{\partial x_i}, \qquad (9)$$

noting that the last term $\partial \Phi / \partial x_i$ did not appear in the corresponding equation of Engels (1975); and differentiate:

$$\frac{\partial \phi_i}{\partial \dot{x}_j} = \frac{\partial g_i}{\partial \dot{x}_j} - \frac{\partial^3 \Phi}{\partial \dot{x}_i \partial \dot{x}_j \partial x_k} \dot{x}_k$$

$$- \frac{\partial^3 \Phi}{\partial \dot{x}_i \partial \dot{x}_j \partial t} - \frac{\partial^2 \Phi}{\partial \dot{x}_i \partial x_j} + \frac{\partial^2 \Phi}{\partial \dot{x}_j \partial x_i},$$

which yields by (8) the following identity:

$$\frac{\partial \phi_i}{\partial \dot{x}_j} + \frac{\partial \phi_j}{\partial \dot{x}_i} = \frac{\partial g_i}{\partial \dot{x}_j} + \frac{\partial g_j}{\partial \dot{x}_i}$$

$$\qquad (10)$$

$$- 2 \left(\frac{\partial^3 \Phi}{\partial \dot{x}_i \partial \dot{x}_j \partial x_k} \dot{x}_k + \frac{\partial^3 \Phi}{\partial \dot{x}_i \partial \dot{x}_j \partial t} \right) = 0.$$

This identity is moreover differentiated and the indices i and k are interchanged to obtain

$$\frac{\partial^2 \phi_i}{\partial \dot{x}_k \partial \dot{x}_j} + \frac{\partial^2 \phi_j}{\partial \dot{x}_k \partial \dot{x}_i} = 0, \qquad \frac{\partial^2 \phi_i}{\partial \dot{x}_j \partial \dot{x}_k} + \frac{\partial^2 \phi_k}{\partial \dot{x}_j \partial \dot{x}_i} = 0;$$

which yield by adding

$$2 \frac{\partial^2 \phi_i}{\partial \dot{x}_k \partial \dot{x}_j} + \frac{\partial}{\partial \dot{x}_i} \left(\frac{\partial \phi_i}{\partial \dot{x}_k} + \frac{\partial \phi_k}{\partial \dot{x}_j} \right) = 2 \frac{\partial^2 \phi_i}{\partial \dot{x}_k \partial \dot{x}_j} = 0.$$

Hence the functions ϕ_i are integrated as

$$\phi_i = F_{ij}(x, t)\, \dot{x}_j + F_i(x, t), \tag{11}$$

in which (10) imposes the condition

$$F_{ij} + F_{ji} = 0. \tag{12}$$

Therefore (9) implies that the functions g_i are of the forms

$$g_i = \frac{\partial^2 \Phi}{\partial \dot{x}_i \partial x_k}\, \dot{x}_k + \frac{\partial^2 \Phi}{\partial \dot{x}_i \partial t} - \frac{\partial \Phi}{\partial x_i} + F_{ij}\, \dot{x}_j + F_i . \tag{13}$$

The functions g_i are put into (6e) with the following differentiations:

$$\frac{\partial g_i}{\partial x_j} = \frac{\partial^3 \Phi}{\partial \dot{x}_i \partial x_j \partial x_k}\, \dot{x}_k + \frac{\partial^3 \Phi}{\partial \dot{x}_i \partial x_j \partial t} + \frac{\partial F_{ik}}{\partial x_j}\, \dot{x}_j + \frac{\partial F_i}{\partial x_j} + (i, j).$$

$$\frac{\partial g_i}{\partial \dot{x}_j} = \frac{\partial^2 \Phi}{\partial \dot{x}_i \partial x_j} - \frac{\partial^2 \Phi}{\partial \dot{x}_j \partial x_i} + F_{ij} + (i, j),$$

$$\frac{\partial^2 g_i}{\partial \dot{x}_j \partial x_k} = \frac{\partial^3 \Phi}{\partial \dot{x}_i \partial x_j \partial x_k} - \frac{\partial^3 \Phi}{\partial \dot{x}_j \partial x_i \partial x_k} + \frac{\partial F_{ij}}{\partial x_k} + (i, j),$$

$$\frac{\partial^2 g_i}{\partial \dot{x}_j \partial t} = \frac{\partial^3 \Phi}{\partial \dot{x}_i \partial x_j \partial t} - \frac{\partial^3 \Phi}{\partial \dot{x}_j \partial x_i \partial t} + \frac{\partial F_{ij}}{\partial t} + (i, j),$$

where (i, j) denotes the remaining symmetric terms with respect to the indices i and j, which vanishes after substitution. Then, since the Φ's terms in both sides of (6e) are cancelled, by using (12) it is concluded:

$$\left(\frac{\partial F_{ij}}{\partial x_k} + \frac{\partial F_{jk}}{\partial x_i} + \frac{\partial F_{ki}}{\partial x_j} \right) \dot{x}_k$$

$$- \left(\frac{\partial F_i}{\partial x_j} - \frac{\partial F_j}{\partial x_i} - \frac{\partial F_{ij}}{\partial t} \right) = 0 ,$$

which is identical for arbitrary x_i if and only if

$$\frac{\partial F_{ij}}{\partial x_k} + \frac{\partial F_{jk}}{\partial x_i} + \frac{\partial F_{ki}}{\partial x_j} = 0 , \tag{14}$$

$$\frac{\partial F_i}{\partial x_j} - \frac{\partial F_j}{\partial x_i} = \frac{\partial F_{ij}}{\partial t} . \tag{15}$$

The conditions (12) and (14) are necessary and sufficient for the existence of functions $G_i (x, t)$, such that

$$F_{ij} = \frac{\partial G_i}{\partial x_j} - \frac{\partial G_j}{\partial x_i} . \tag{16}$$

This is put into (15) to derive the condition

$$\frac{\partial}{\partial x_j} \left(\frac{\partial G_i}{\partial t} - F_i \right) = \frac{\partial}{\partial x_i} \left(\frac{\partial G_j}{\partial t} - F_j \right) ,$$

which guarantees the existence of a function $F(x, t)$, such that

$$\frac{\partial G_i}{\partial t} - F_i = \frac{\partial F}{\partial x_i} , \text{ i.e., } F_i = - \frac{\partial F}{\partial x_i} + \frac{\partial G_i}{\partial t} . \tag{17}$$

Thus (16) and (17) determine ϕ_i of (11) as follows:

$$\phi_i = \left(\frac{\partial G_i}{\partial x_j} - \frac{\partial G_j}{\partial x_i} \right) \dot{x}_j - \frac{\partial F}{\partial x_i} + \frac{\partial G_i}{\partial t},$$

and g_i of (13) also

$$g_i = \frac{\partial^2 \Phi}{\partial \dot{x}_i \partial x_k} \dot{x}_k + \frac{\partial^2 \Phi}{\partial \dot{x}_i \partial t} - \frac{\partial \Phi}{\partial x_i}$$

$$+ \left(\frac{\partial G_i}{\partial x_j} - \frac{\partial G_j}{\partial x_i} \right) \dot{x}_j - \frac{\partial F}{\partial x_i} + \frac{\partial G_i}{\partial t}. \quad (18)$$

We are now in the position to construct the Lagrangean under consideration. Define a function $\phi (\dot{x}, x, t)$ by

$$\phi = G_i (x, t) \dot{x}_i + F(x, t), \quad (19)$$

then from (7) if follows that

$$G_{ij} = \frac{\partial^2 \Phi}{\partial \dot{x}_i \partial \dot{x}_j} = \frac{\partial^2 (\Phi + \phi)}{\partial \dot{x}_i \partial \dot{x}_j}. \quad (20)$$

The following identify obtained by direct calculation of (19):

$$\frac{\partial^2 \phi}{\partial \dot{x}_i \partial x_j} \dot{x}_j + \frac{\partial^2 \phi}{\partial \dot{x}_i \partial t} - \frac{\partial \phi}{\partial x_i}$$

$$= \left(\frac{\partial G_i}{\partial x_j} - \frac{\partial G_j}{\partial x_i} \right) \dot{x}_j - \frac{\partial F}{\partial x_i} + \frac{\partial G_i}{\partial t} = \phi_i$$

is used to rewrite g_i of (18) as

$$g_i = \frac{\partial^2 (\Phi + \phi)}{\partial \dot{x}_i \partial x_k} \dot{x}_k + \frac{\partial^2 (\Phi + \phi)}{\partial \dot{x}_i \partial t} - \frac{\partial (\Phi + \phi)}{\partial x_i}. \quad (21)$$

Equations (20) and (21) imply that the middle term of differential equations (3) are just identified with $[\Phi + \phi]_i$, i.e.,

$$G_{ij} \ddot{x}_j + g_i = [\Phi + \phi]_i .$$

Thus, for the given system of differential equations (3), the Lagrangian in the inverse problem is constructed as

$$\Phi + \phi = \Phi(\dot{x}, x, t) + G_i(x, t) \dot{x}_i + F(x, t).$$

This completes the inverse proof of the Helmholtz conditions.

III. THE DERIVATION OF CONSERVATION LAWS WITH THE AID OF HELMHOLTZ CONDITIONS

For the case $n = 1$, a procedure for deriving conservation laws, including those that cannot be obtained from the Noether symmetries, can be constructed with the aid of the Helmholtz conditions. In this case, (3) becomes the single differential equation

$$G(\dot{x}, x, t) \ddot{x} + g(\dot{x}, x, t) = 0,$$

and whose Helmholtz conditions corresponding to (6) are always satisfied with the exception of condition (6c):

$$\frac{\partial g}{\partial x} = \frac{\partial G}{\partial x} \dot{x} + \frac{\partial G}{\partial t}. \quad (22)$$

For a given Lagrangean $L(\dot{x}, x, t)$, if there exists a new Lagrangean $N(\dot{x}, x, t)$, such that [see equation (1).]

$$[N] = C(\dot{x}, x, t) [L],$$

i.e., $C[L] = 0$ is identified with some Euler-Lagrange equation $[N] = 0$; then C is a constant of the motion: $C = $ const. (conservation law) of the Euler-Lagrange equation $[L] = 0$. Since $C[L] = CK\ddot{x} - Ck$ where

$$K = \frac{\partial^2 L}{\partial \dot{x}^2}, \quad k = \frac{\partial^2 L}{\partial \dot{x}\, \partial x} \dot{x} + \frac{\partial^2 L}{\partial \dot{x}\, \partial t} - \frac{\partial L}{\partial x}, \qquad (23)$$

the Helmholtz condition (22) is translated into

$$-\frac{\partial (Ck)}{\partial \dot{x}} = \dot{x}\, \frac{\partial (CK)}{\partial x} + \frac{\partial (CK)}{\partial t}, \qquad \qquad \text{i.e.,}$$

$$k \frac{\partial C}{\partial \dot{x}} + \dot{x} K \frac{\partial C}{\partial x} + K \frac{\partial C}{\partial t} + \left(\frac{\partial k}{\partial \dot{x}} + \dot{x} \frac{\partial K}{\partial x} + \frac{\partial K}{\partial t} \right) C = 0.$$

By putting (23) into the above equation, the terms $\partial k / \partial \dot{x} + \dot{x}\, \partial K / \partial x + \partial K / \partial t$ vanish and the remaining terms are rewritten as

$$\left(\dot{x} \frac{\partial^2 L}{\partial \dot{x}\, \partial x} + \frac{\partial^2 L}{\partial \dot{x}\, \partial t} - \frac{\partial L}{\partial x} \right) \frac{\partial C}{\partial \dot{x}}$$

$$- \dot{x} \frac{\partial^2 L}{\partial \dot{x}^2} \frac{\partial C}{\partial x} - \frac{\partial^2 L}{\partial \dot{x}^2} \frac{\partial C}{\partial t} = 0. \qquad (24)$$

By a particular Lagrangean of the form

$$L = e^{-\rho t} F(\dot{x}, x), \qquad \rho = \text{const.},$$

with the Euler-Lagrange equation

$$\ddot{x} \, \frac{\partial^2 F}{\partial \dot{x}^2} + \dot{x} \, \frac{\partial^2 F}{\partial \dot{x} \, \partial t} - \rho \, \frac{\partial F}{\partial \dot{x}} - \frac{\partial F}{\partial x} = 0; \qquad (25)$$

the equation (24) is moreover rewritten as

$$\left(\dot{x} \, \frac{\partial^2 F}{\partial \dot{x} \, \partial x} - \rho \, \frac{\partial F}{\partial \dot{x}} - \frac{\partial F}{\partial x} \right) \frac{\partial C}{\partial \dot{x}}$$

$$- \dot{x} \, \frac{\partial^2 F}{\partial \dot{x}^2} \, \frac{\partial C}{\partial x} - \frac{\partial^2 F}{\partial \dot{x}^2} \, \frac{\partial C}{\partial t} = 0. \quad (26)$$

Thus the following theorem is obtained for deriving conservation laws.

Theorem 1. *For a given Lagrangean $L(\dot{x}, x, t)$ or particularly of the form $L = e^{-\rho t} F(\dot{x}, x)$, $\rho = $ const., any solution $C(\dot{x}, x, t)$ of the respective differential equation (24) or (26) is a constant of the motion, i.e., the solution leads a conservation law $C(\dot{x}, x, t) = $ const. of the Euler-Lagrange equation $[L] = 0$.*

IV. APPLICATION OF METHOD TO NEOCLASSICAL GROWTH MODELS

Theorem 1 is now applied to Neoclassical optimal growth models to obtain a class of conservation laws, including non-Noether ones in the models, while the Noether types were completed by Sato (1981) under the group theoretic considerations of symmetries.

1. Simple model of the Ramsey type (1928). In this case, the Lagrangean $L(\dot{x}, x, t)$ takes the form

$$L = e^{-\rho t} U(c), \qquad c = g(x) - \dot{x}, \qquad (27)$$

where U is the utility (or welfare) function of the consumption c with the usual properties $U' > 0$ and $U'' < 0$, x the capital-labor ratio, $\dot{x}$ the rate of capital accumulation and $\rho \geq 0$ the fixed discount rate. With the Lagrangian L of the form (27), the Euler-Lagrange equation (25) is reduced to

$$(\dot{x}g' - \ddot{x}) U'' + (g' - \rho) U' = 0; \qquad (28)$$

and the differential equation (26) is also to

$$\left(\dot{x}g' \frac{\partial C}{\partial \dot{x}} + \dot{x} \frac{\partial C}{\partial x} + \frac{\partial C}{\partial t} \right) U'' + (g' - \rho) \frac{\partial C}{\partial \dot{x}} U' = 0, \qquad (29)$$

in solutions of which, a class of conservation laws of (21) can be discovered.

Case 1: $\rho = 0$. With an assumption $C = C(\dot{x}, x)$, by using a new function

$$V = U + \dot{x}U', \qquad (30)$$

the equation (29) is rewritten as

$$\frac{\partial V}{\partial \dot{x}} \frac{\partial C}{\partial x} - \frac{\partial V}{\partial \dot{x}} \frac{\partial C}{\partial x} = 0, \tag{31}$$

whose subsidiary equation

$$\frac{d\dot{x}}{\partial V / \partial x} = - \frac{dx}{\partial V / \partial \dot{x}}, \qquad \text{i.e., } dV = 0$$

has a solution $V = $ const. Hence the solution $C(\dot{x}, x)$ of (31) is determined as

$$C = \Phi (U + \dot{x}U'). \tag{32}$$

Case 2: $\rho \neq 0$. To derive (global) conservation laws, assume that (29) is satisfied for arbitrary utility functions U, i.e.,

$$\dot{x}g' \frac{\partial C}{\partial \dot{x}} + \dot{x} \frac{\partial C}{\partial x} + \frac{\partial C}{\partial t} = 0, \tag{33}$$

$$(g' - \rho) \frac{\partial C}{\partial \dot{x}} = 0. \tag{34}$$

In (33), $\partial C / \partial \dot{x} = 0$ or equivalently $C = C(x, t)$ implies $\dot{x} \partial C / \partial x + \partial C / \partial t = 0$, i.e., $\partial C / \partial x = \partial C / \partial t = 0$; hence $C = $ const. So let $C \neq $ const., i.e., $\partial C / \partial \dot{x} \neq 0$. Then (34) yields

$$g' = \rho, \quad \text{i.e.,} \quad g = \rho x + \beta, \quad \beta = \text{const.};$$

which makes (33) into the form

$$\rho \dot{x} \frac{\partial C}{\partial \dot{x}} + \dot{x} \frac{\partial C}{\partial x} + \frac{\partial C}{\partial t} = 0. \tag{35}$$

with the subsidiary equation

$$\frac{d\dot{x}}{\rho\dot{x}} = \frac{dx}{\dot{x}} = \frac{dt}{1}$$

The first equation: $d\dot{x} / \rho\dot{x} = d\dot{x} / x$, i.e., $d\dot{x} = \rho \, dx$ implies

$$- \dot{x} + \rho x = k = \text{const.,} \qquad (36)$$

$$\text{i.e.,} \quad \dot{x} = \rho x - k;$$

and hence the second one becomes

$$\frac{\rho \, dx}{\rho \, x - k} = \frac{\rho \, dt}{1},$$

which is integrated as

$$\log | \rho x - k | - \rho t = \text{const., i.e.,}$$

$$e^{-\rho t} (\rho x - k) = e^{-\rho t} \dot{x} = \text{const.} \qquad (37)$$

Hence, the independent solutions (36) and (37) determine the solution of (35) as follows:

$$C = \Phi(e^{-\rho t} \dot{x}, - \dot{x} + \rho x). \qquad (38)$$

Thus, the following class of conservation laws is discovered in the solutions (32) and (38). In what follows, the conservation laws that have appeared in Sato (1981) and the others (incuding non-Noether type) are distinguished by Greek letters Ω and Ξ, respectively.

Theorem 2. *(Conservation laws for simple models of the Ramsey type.) When the discount rate ρ is zero, there exists only one conservation law:*

$$\Omega = U(c) + \dot{x}U'(c) = const.$$

When the discount rate ρ is positive, there exists no (global) conservation law unless the case that the consumption is of the form

$$c = -\dot{x} + \rho x + \beta, \qquad \beta = const.;$$

and in the case there exist two independent conservation laws,

$$\Omega = -\dot{x} + \rho x = const., i.e., c = const.,$$
$$\Xi = e^{-\rho t}\dot{x} = const.$$

2. The Liviatan-Samuelson model (1969). The Lagrangean L of the form (27) can be generalized in the Liviatan-Samuelson model:

$$L = e^{-\rho t} U(c), \quad c = f(\dot{x}, x),$$

where f is the transformation function (production possibility frontier) with the usual concavity to the origin. Hence $\partial^2 f / \partial \dot{x}^2 < 0$, while $\partial^2 f / \partial \dot{x}^2 = 0$ in the Ramsey type of 1. This Lagrangean translates the Euler-Lagrange equation (25) into

$$\left(x \left(\frac{\partial f}{\partial \dot{x}} \right)^2 + \dot{x} \frac{\partial f}{\partial \dot{x}} \frac{\partial f}{\partial x} \right) U''$$

$$+ \left(\ddot{x} \frac{\partial^2 f}{\partial \dot{x}^2} + \dot{x} \frac{\partial^2 f}{\partial \dot{x} \partial x} - \rho \frac{\partial f}{\partial \dot{x}} - \frac{\partial f}{\partial x} \right) U' = 0,$$

and also the differential equation (26) into

$$
\left(\dot{x} \, \frac{\partial f}{\partial \dot{x}} \, \frac{\partial f}{\partial x} \, \frac{\partial C}{\partial \dot{x}} - \left(\frac{\partial f}{\partial \dot{x}} \right)^2 \left(\dot{x} \, \frac{\partial C}{\partial x} + \frac{\partial C}{\partial t} \right) \right) U''
$$

$$
+ \left(\left(\dot{x} \, \frac{\partial^2 f}{\partial \dot{x} \, \partial x} - \rho \frac{\partial f}{\partial \dot{x}} - \frac{\partial f}{\partial x} \right) \frac{\partial C}{\partial \dot{x}} \right.
$$

$$
\left. - \frac{\partial^2 f}{\partial \dot{x}^2} \left(\dot{x} \, \frac{\partial C}{\partial x} + \frac{\partial c}{\partial t} \right) \right) U' = 0. \tag{39}
$$

Case 1: $\rho = 0$. With the assumption $C = C(\dot{x}, x)$, this case does not differ from case 1 of the Ramsey model, but the function V of (30) is replaced with a generalization of (30):

$$
V = U - \dot{x} \, \frac{\partial f}{\partial \dot{x}} \, U'.
$$

Hence the solution of (39) is determined similarly as

$$
C = \Phi \left(U - \dot{x} \, \frac{\partial f}{\partial \dot{x}} \, U' \right). \tag{40}
$$

Case 2: $\rho \neq 0$. The equation (39) is satisfied for arbitrary utility functions U if and only if (note $\partial f / \partial \dot{x} < 0$ follows from the concavity of f, so that $\partial f / \partial \dot{x} \neq 0$)

$$\dot{x} \frac{\partial f}{\partial x} \frac{\partial C}{\partial \dot{x}} - \frac{\partial f}{\partial x} \left(\dot{x} \frac{\partial C}{\partial x} + \frac{\partial C}{\partial t} \right) = 0, \ (41)$$

$$\left(\dot{x} \frac{\partial^2 f}{\partial \dot{x} \partial x} - \rho \frac{\partial f}{\partial \dot{x}} - \frac{\partial f}{\partial x} \right) \frac{\partial C}{\partial \dot{x}} - \frac{\partial^2 f}{\partial \dot{x}^2} \left(\dot{x} \frac{\partial C}{\partial x} + \frac{\partial C}{\partial t} \right) = 0. \ (42)$$

In these equations, assume that $\partial C / \partial \dot{x} \neq 0$ or $\dot{x} \partial C / \partial x + \partial C / \partial t \neq 0$, i.e., $C \neq$ const. to obtain

$$\dot{x} \frac{\partial^2 f}{\partial \dot{x}^2} \frac{\partial f}{\partial x} = \frac{\partial f}{\partial \dot{x}} \left(\dot{x} \frac{\partial^2 f}{\partial \dot{x} \partial x} - \rho \frac{\partial f}{\partial \dot{x}} - \frac{\partial f}{\partial x} \right), \qquad \text{i.e.,}$$

$$\frac{\partial f / \partial x}{\partial f / \partial \dot{x}} + \rho = \dot{x} \frac{(\partial^2 f / \partial \dot{x} \partial x)(\partial f / \partial \dot{x}) - (\partial f / \partial x)(\partial^2 f / \partial \dot{x}^2)}{(\partial f / \partial \dot{x})^2}$$

$$= \dot{x} \frac{\partial}{\partial \dot{x}} \left(\frac{\partial f / \partial x}{\partial f / \partial \dot{x}} \right).$$

Hence by putting

$$\phi = \frac{\partial f / \partial x}{\partial f / \partial \dot{x}}, \qquad \text{i.e.,} \quad \phi \frac{\partial f}{\partial \dot{x}} - \frac{\partial f}{\partial x} = 0, \qquad (43)$$

the following differential equation is derived:

$$\dot{x} \frac{\partial \phi}{\partial \dot{x}} = \phi + \rho. \tag{44}$$

If $\phi + \rho = 0$, the equation (43) implies $\rho \, \partial f / \partial \dot{x} + \partial f / \partial x = 0$, i.e., $f = f(-\dot{x} + \rho \, x)$; which may be written as $f = f(-\dot{x} + \rho \, x + \beta)$, where $c = -\dot{x} + \rho x + \beta$ is the consumption in the case 2 of the Ramsey model. Since, with this transformation function f, the equation (41), or equivalently (42), yields the same differential equation (35), similarly

$$C = \Phi \, (e^{-\rho t} \, \dot{x}, -\dot{x} + \rho \, x) \qquad (38)$$

is a solution of (41) or (42). If $\phi + \rho \neq 0$, (44) is rewritten as

$$\frac{1}{\phi + \rho} \, \frac{\partial \phi}{\partial \dot{x}} - \frac{1}{\dot{x}} = 0, \quad \text{i.e.,} \quad \frac{\partial}{\partial \dot{x}} \left(\log | \, \phi + \rho \, | - \log | \, \dot{x} \, | \right) = 0,$$

which is integrated to determine ϕ :

$$\phi = e^{h(x)} \, \dot{x} - \rho.$$

In this case, (43) becomes

$$(e^h \, \dot{x} - \rho) \, \frac{\partial f}{\partial \dot{x}} - \frac{\partial f}{\partial x} = 0, \qquad (45)$$

with subsidiary equation

$$d\dot{x} + (e^h \, \dot{x} - \rho) \, dx = 0 \quad \text{or} \quad e^p \, d\dot{x} + e^p(e^h \, \dot{x} - \rho) \, dx = 0, \quad (46)$$

where $p = p(x)$ is the integrating factor

$$p(x) = \int e^{h(x)} \, dx.$$

Hence the transformation function f is determined as

$$f = f(-e^p \dot{x} + \rho \int e^p \, dx),$$

which translates the differential equation (41) or (42) into

$$\dot{x}(e^h \dot{x} - \rho) \frac{\partial C}{\partial \dot{x}} - \dot{x} \frac{\partial C}{\partial x} - \frac{\partial C}{\partial t} = 0, \qquad (47)$$

with subsidiary equation

$$\frac{d\dot{x}}{\dot{x}(e^h \dot{x} - \rho)} = \frac{dx}{-\dot{x}} = \frac{dt}{-1}.$$

Since the first equation above does not differ from (46), the similar solution is obtained:

$$-e^p \dot{x} + \rho \int e^p \, dx = k = \text{const.,} \qquad (48)$$

i.e., $\dot{x} = e^{-p}(\rho \int e^p \, dx - k);$

and hence the second one becomes

$$\frac{\rho \, e^p dx}{\rho \int e^p \, dx - k} = \frac{\rho \, dt}{1},$$

which is integrated as

$$\log | \rho \int e^p \, dx - k | - \rho \, t = \text{const.,} \quad \text{i.e.,}$$

$$e^{-pt}(\rho \int e^p \, dx - k) = e^{-pt+p} \dot{x} = \text{const.} \qquad (49)$$

Hence, the independent solutions (48) and (49) determine the solution of (47) as follows:

$$C = \Phi(e^{-pt+p} \dot{x}, -e^p \dot{x} + \rho \int e^p \, dx). \qquad (50)$$

Thus the following conservation laws are discovered in the solutions (40), (38) and (50).

Theorem 3. *(The Liviatan-Samuelson models.) When the discount rate ρ is zero, there exists only one conservation law:*

$$\Xi = U(c) - \dot{x} \frac{\partial f}{\partial \dot{x}} U'(c) = const.$$

When the discount rate ρ is positive, there exists no (global) conservation law unless the case that the consumption is of the form

(i) $c = f(-\dot{x} + \rho x + \beta)$, $\beta = const.$,

or, (ii) $c = f(-e^p \dot{x} + \rho \int e^p dx)$, $p = \int e^{h(x)} dx$.

In case (i), there exist two independent conservation laws:

$$\Xi = -\dot{x} + \rho x = const., \quad i.e., c = const.,$$

$$\Xi = e^{-\rho t} \dot{x} = const.;$$

or in case (ii), there exist two independent conservation laws:

$$\Xi = -e^p \dot{x} + \rho \int e^p dx = const., \quad i.e., c = const.,$$

$$\Xi = e^{-\rho t + p} \dot{x} = const.$$

3. The model suggested by Samuelson (1972). The final example is for the Lagrangean $L(\dot{x}, x, t)$ suggested by Samuelson:

$$L = e^{-\rho t} \left(-\frac{1}{2} \dot{x}^2 - a\dot{x}x - \frac{1}{2} x^2\right), \quad -1 < a < 1,$$

with the Euler-Lagrange equation

$$\ddot{x} - \rho \dot{x} - (1 + a\rho) x = 0.$$

In terms of this Lagrangean, the differential equation (26) is reduced to

$$(\rho \dot{x} + (1 + a\rho) x) \frac{\partial C}{\partial \dot{x}} + \dot{x} \frac{\partial C}{\partial x} + \frac{\partial C}{\partial t} = 0. \tag{51}$$

Now, into the above equation, take the transformation

$$\begin{pmatrix} \dot{y} \\ y \end{pmatrix} = P \begin{pmatrix} \dot{x} \\ x \end{pmatrix}, \quad P = \begin{pmatrix} 1 & -(1/2)(r + \rho) \\ 1 & (1/2)(r - \rho) \end{pmatrix},$$

where $r = \sqrt{\rho^2 + 4\sigma}$ and $\sigma = 1 + a\rho$, so that

$$r^2 = (\rho + 2a)^2 + 4(1 - a^2) > 0, \quad \text{i.e.,} \quad \det P = r > 0,$$

i.e., the transformation matrix P is nonsingular. Since the change of differentiating variables implies

$$\left(\frac{\partial C}{\partial \dot{x}} \; \frac{\partial C}{\partial x} \right) = \left(\frac{\partial C}{\partial \dot{y}} \; \frac{\partial C}{\partial y} \right) \begin{pmatrix} \dfrac{\partial \dot{y}}{\partial \dot{x}} & \dfrac{\partial \dot{y}}{\partial x} \\ \dfrac{\partial y}{\partial \dot{x}} & \dfrac{\partial y}{\partial x} \end{pmatrix} = \left(\frac{\partial C}{\partial \dot{y}} \; \frac{\partial C}{\partial y} \right) P,$$

the terms in (51) is transformed to

$$(\rho \dot{x} + (1 + a\rho) x) \frac{\partial C}{\partial \dot{x}} + \dot{x} \frac{\partial C}{\partial x}$$

$$= \left(\frac{\partial C}{\partial \dot{x}} \; \frac{\partial C}{\partial x} \right) \begin{pmatrix} \rho & \sigma \\ 1 & 0 \end{pmatrix} \begin{pmatrix} \dot{x} \\ x \end{pmatrix}$$

$$= \left(\frac{\partial C}{\partial \dot{y}} \; \frac{\partial C}{\partial y} \right) P \begin{pmatrix} \rho & \sigma \\ 1 & 0 \end{pmatrix} P^{-1} \begin{pmatrix} \dot{y} \\ y \end{pmatrix};$$

in which the identity holds:

$$
P \begin{pmatrix} \rho & \sigma \\ 1 & 0 \end{pmatrix} P^{-1} = \begin{pmatrix} (1/2)\,(\rho - r) & 0 \\ 0 & (1/2)\,(\rho + r) \end{pmatrix}.
$$

Hence the equation (51) is transformed to

$$
\alpha \dot{y}\, \frac{\partial C}{\partial \dot{y}} + \beta y\, \frac{\partial C}{\partial y} + \frac{\partial C}{\partial t} = 0; \qquad (52)
$$

whose subsidiary equation

$$
\frac{d\dot{y}}{\alpha \dot{y}} = \frac{dy}{\beta y} = \frac{dt}{1}
$$

has independent solutions:

$$
\dot{y}^{\beta} y^{-\alpha} = \text{const., i.e.,} \quad \dot{y}^{2\beta} y^{-2\alpha} = \text{const.;} \quad e^{-\alpha t}\, \dot{y} = \text{const.,} \qquad (53)
$$

or
$$
\dot{y}^{2\beta} y^{-2\alpha} = \text{const.,} \quad e^{-\beta t}\, y = \text{const.,} \qquad (54)
$$

where $\alpha = (1/2)\,(\rho - r)$ and $\beta = (1/2)\,(\rho + r)$. Corresponding to the independent solutions (53) or (54), the solution of (52), and hence the final solution of (51), are determined respectively as follows:

$$
C = \Phi\,(e^{-\alpha t}\, \dot{y},\ \dot{y}^{2\beta} y^{-2\alpha})
$$

$$
= \Phi \left[e^{(r-\rho)\,t/2} \left(\dot{x} - \frac{r+\rho}{2}\, x \right),\ \left(\dot{x} - \frac{r+\rho}{2}\, x \right)^{r+\rho} \left(\dot{x} + \frac{r-\rho}{2}\, x \right)^{r-\rho} \right],
$$

or

$$C = \Phi \left(e^{-\beta t} y, \; \dot{y}^{2\beta} y^{-2\alpha} \right)$$

$$= \Phi \left[e^{-(r-\rho)\,t/2} \left(\dot{x} - \frac{r+\rho}{2} x \right), \; \left(\dot{x} - \frac{r+\rho}{2} x \right)^{r+\rho} \left(\dot{x} + \frac{r-\rho}{2} x \right)^{r-\rho} \right].$$

Thus the following conservation laws are discovered in the above solutions Φ.

Theorem 4. *(The model suggested by Samuelson: $L = e^{-\rho t} (-(1/2) \dot{x}^2 - a\dot{x}x - (1/2)x^2), -1 < a < 1$). In this model, there exist the following three conservation laws:*

$$\Omega_4 = e^{(r-\rho)\,t/2} \left(\dot{x} - \frac{r+\rho}{2} x \right) = const.,$$

$$\Omega_5 = e^{-(r+\rho)\,t/2} \left(\dot{x} - \frac{r-\rho}{2} x \right) = const.,$$

$$\Xi = \left(\dot{x} - \frac{r+\rho}{2} x \right)^{r+\rho} \left(\dot{x} + \frac{r-\rho}{2} x \right)^{r-\rho} = const.,$$

in which Ω_4 and Ω_5 are independent and Ξ is dependent on Ω_4 and Ω_5, i.e., $\Xi = \Omega_4^{r+\rho} \Omega_5^{r-\rho}$.

Remark. The subnumbers of Ω coincide with those obtained by Sato (1981). The others Ω_1, Ω_2 and Ω_3 in Sato (1981) are dependent on Ω_4 and Ω_5 (the statement of independence of Ω_3, Ω_4 and Ω_5 in Sato (1981) is not correct: $\Omega_1 = \Omega_4^2$, $\Omega_2 = \Omega_5^2$ and $\Omega_3 = (1/2)\,\Omega_4 \times \Omega_5$. Also, Ω_1 and Ω_2 can be discovered in our method. In fact, since the independent solutions (53) or (54) of (52) may be rewritten as

$$\dot{y}^{2\beta} y^{-2\alpha} = const., \quad e^{-2\alpha t} \dot{y}^2 = const., \text{ or}$$

$$\dot{y}^{2\beta} y^{-2\alpha} = const., \quad e^{-2\beta t} \dot{y}^2 = const.,$$

and moreover, the following are independent solutions of (52):

$$e^{-2\alpha t}\dot{y}^2 = \text{const.}, \quad e^{-2\beta t}y^2 = \text{const.};$$

another expression of the solution of (51) can be obtained as follows respectively

$$C = \Phi(e^{-2\alpha t}\dot{y}^2, \, \dot{y}^{2\beta}y^{-2\alpha}) = \Phi(\Omega_1, \Xi)$$

$$= \Phi\left[e^{(r-\rho)t}\left(\dot{x} - \frac{r+\rho}{2}x\right)^2, \left(\dot{x} - \frac{r+\rho}{2}x\right)^{r+\rho}\left(\dot{x} + \frac{r-\rho}{2}x\right)^{r-\rho}\right],$$

$$C = \Phi(e^{-2\beta t}y^2, \, \dot{y}^{2\beta}y^{-2\alpha}) = \Phi(\Omega_2, \Xi)$$

$$= \Phi\left[e^{-(r-\rho)t}\left(\dot{x} - \frac{r+\rho}{2}x\right)^2, \left(\dot{x} - \frac{r+\rho}{2}x\right)^{r+\rho}\left(\dot{x} + \frac{r-\rho}{2}x\right)^{r-\rho}\right],$$

$$C = \Phi(e^{-2\alpha t}y^2, \, y^{2\beta}y^2) = \Phi(\Omega_1, \Omega_2)$$

$$= \Phi\left[e^{(r-\rho)t}\left(x - \frac{r+\rho}{2}x\right)^2, \, e^{(r+\rho)t}\left(\dot{x} - \frac{r+\rho}{2}x\right)^2\right].$$

V. CONCLUSION

In this chapter we have discussed a method whereby conservation laws can be derived with the aid of Helmholtz conditions. The procedure, for the case where $n = 1$, was outlined in Section **III**, and then applied to three Neoclassical economic growth models. The results indicate that the method introduced in this chapter represents a useful way of discovering conservation laws in dynamic economic models. The approach enables one to discover non-Noether type as well as Noether type conservation laws.

REFERENCES

Boehm, K. (1900). Die Existenzbedingungen eines von den ersten und zweiten Differentialguotienten der Coordinaten abhangigen kinetischen Potentials. *J. Reine Angew. Math.*, *121*, 124-140.

Crampin, M. (1981). On the differential geometry of the Euler-Lagrange equations, and the inverse problem of Lagrangian dynamics. *J. Phys. A: Math. Gen.*, *14*, 2567-2575.

Currie, D. G. and Saletan, E. J. (1966). *q*-Equivalent particle Hamiltonians – I: The classical one-dimensional case. *J. Math. Phys.*, *7*, 967-974

Darboux, G. (1894). *Leçons sur la théorie générale des surfaces.* Paris: Gauthier-Villars; also (1972) New York: Chelsea.

Davis, D. R. (1928). The inverse problem of the calculus of variations in higher space. *Trans. Am. Math. Soc.*, *30*, 710-736.

Douglas, J. (1941). Solutions of the inverse problem of the calculus of variations. *Trans. Am. Math. Soc.*, *50*, 71-128.

Engels, E. (1975). On the Helmholtz conditions for the existence of a Lagrangean formalism. *Nuovo Cimento*, *26B*, 481-492.

Farrias, J. R. and Nergi, L. J. (1983). Another demonstration of the theory by Hojman and Harleston. *J. Phys. A: Math. Gen.*, *16*, 707-710.

Farrias, J. R. and Teixeira, N. L. (1983). Equivalent Lagrangian in field theory. *J. Phys. A: Math. Gen.*, *16*, 1517-1581.

González-Gascón, F. (1982). New results on first integrals of Lagrangean systems. *Phys. Lett.*, *87A*, 385-386.

Hamel, G. (1903). Über die Geometrieen in denen die Geraden die Kürzeste sind. *Math. Ann.*, *57*, 231-264.

Havas, P. (1957). The range of application of the Lagrange formalism – I. *Suppl. Nuovo Cimento*, *5*, 363-388.

Helmholtz, H. (1887). Über die physikalische Bedeutung des Princips der Kleinsten Wirkung. *J. Reine Angew. Math.*, *100*, 137-166.

Henneaux, M. (1981). On a theorem by Hojman and Harleston. *Hadronic J.*, *4*, 2137-2143.

Henneaux, M. (1982a). Equation of motion, commutation relations and ambiguities in the Lagrangian formalism. *Ann. Phys.*, *140*, 45-64.

Henneaux, M. (1982b). On the inverse problem of the calculus of variations. *J. Phys. A: Math. Gen.*, *15*, L93-L96.

Hirsch, A. (1898). Die Existenzbedingungen des verallgemeinerten kinetischen Potentials. *Math. Ann.*, *50*, 429-441.

Hojman, S. and Harleston, H. (1981). Equivalent Lagrangeans: Multidimensional case. *J. Math. Phys.*, *22*, 1414-1419.

Hojman, S. and Urrutia, L. F. (1981). On the inverse problem of the calculus of variations. *J. Math. Phys., 22,* 1896-1903.

Hojman, S. (1984). Symmetries of Lagrangians and of their equations of motion. *J. Phys. A: Math. Gen., 17,* 2399-2412.

Königsberger, L. (1901). *Die Principien der Mechanik.* Leipzig: Teubner.

Kürschak, J. (1905). Über eine charakteristische Eigenschaft, der Differentialgleichungen der Variations rechnung. *Math. Ann., 60,* 157-165.

Liviatan, N. and Samuelson, P. A. (1969). Notes on turnpikes: Stable and unstable. *J. Economic Theory, 1,* 454-475.

Lutzky, M. (1979a). Origin of non-Noether invariants. *Phys. Lett., 75A,* 8-10.

Lutzky, M. (1979b). Non-invariance symmetries and constant of the motion. *Phys. Lett., 72A,* 86-88.

Mayer, A. (1896). Die Existenzbedingungen eines kineteschen Potentiales: Beweis eines Satzes von Helmholtz. *Ber. Ges. Leipzig, Phys., C1,* 519-529.

Mimura, F. and Nôno, T. (1984). Conservation laws derived from equivalent Lagrangians and Hamiltonians in particle dynamics. *Bull. Kyushu Inst. Tech. Math. Natur. Sci., 31,* 27-37.

Mimura, F. and Nôno, T. (1985). Equivalent Lagrangean densities in continuum mechanics associated with dynamical symmetries. *Bull. Kyushu Inst. Tech. Math. Natur. Sci., 32,* 15-30.

Mimura, F. and Nôno, T. (1986). Conservation laws derived from equivalent Lagrangean densities in continuum mechanics. *Bull. Kyushu Inst. Tech. Math. Natur. Sci., 33,* 21-35.

Noether, E. (1918). Invariante Variations problem. *Nachr. Kgl. Ges. Wiss. Göttingen Math. Phys. Kl., II,* 235-257.

Ramsey, F. (1928). A mathematical theory of saving. *Economic J., 38,* 543-559.

Samuelson, P. A. (1972). The general saddlepoint property of optimal-control motions. *J. Economic Theory, 5,* 102-120.

Santilli, R. M. (1977a). Necessary and sufficient conditions for the existence of a Lagrangean in field theory I: Variational approach to self-adjointness for tensorial field equations. *Ann. Phys., 103,* 354-408.

Santilli, R. M. (1977b). Necessary and sufficient conditions for the existence of a Lagrangian in field theory II: Direct analytic representations of tensorial field equations. *Ann. Phys., 103,* 409-468.

Santilli, R. M. (1978). *Foundation of theoretical mechanics I: The inverse problem in Newtonian mechanics.* New York: Springer.

Sarlet, W. (1981). Symmetries, first integrals and the inverse problem of Lagrangean mechanics. *J. Phys. A: Math. Gen., 14,* 2227-2238.

Sarlet, W. (1982). The Helmholtz conditions revisted: A new approach to the inverse problem of Lagrangean dynamics. *J. Phys. A: Math. Gen., 14,* 1503-1517.

Sarlet, W. and Cantrijn, F. (1983). Symmetries, first integrals, and the inverse problem of Lagrangean mechanics – II. *J. Phys. A: Math. Gen., 16*, 1383-1396.

Sato, R. (1981). *Theory of technical change and economic invariance: Application of Lie groups.* New York: Academic Press.

CHAPTER 6

Conservation Laws in Continuous and Discrete Models*

In memory of Professor Mineo Ikeda

Ryuzo Sato
Shigeru Maeda

1. INTRODUCTION

The study of economic conservation laws is still in its infancy relative to its counterparts in physics and engineering. Yet this is an area where there is great interest and rapid progress is being made. In economics, the conservation law has its roots in the most celebrated article of Frank Ramsey (1928). But it was Paul A. Samuelson (1970) who first explicitly introduced the concept of conservation law to theoretical economics. The recent works by Weitzman (1976), Sato (1981, 1985), Kemp and Long (1982), Samuelson (1971, 1982), Sato, Nono and Mimura (1984), and Sato and Maeda (1987) provide an indication of the rapid progress being made in this field.

The main purposes of this paper are to subdivide the known universe on continuous dynamic models into five basic types with the help of the Noether theorem (Sato, 1981) and to introduce a new method of analyzing discrete economic models. The application of this theorem has enabled researchers to uncover many "hidden"

---

* Paper presented at the New York University Japan - U.S. Symposium on "Lie Groups and Related Dynamic Models: Applications to Economics and Finance," May 20, 1987.

conservation laws in physics and engineering and has also been instrumental for the discovery of some unknown invariances in economic models.

Discrete models have played an important role in economic analysis. In fact, some economists consider discrete models to be more realistic and more suitable for empirical applications than continuous models. In economics, continuous models serve as approximations of discrete models, as economic data is almost always measured in discrete time such as in days, weeks, months and years. The second section of this paper will be devoted to the study of economic conservation laws in discrete-time optimal growth models. As the methodology and results are new, we devoted a relatively large portion of the paper to the analysis of discrete models.

It will be shown that there exist several conservation laws including the one very similar to the continuous case. But, the discrete models offer unique invariants and conservatives unknown in the continuous case. There has to be always "adjustment factors" which will modify the standard conservation laws.

2. CONTINUOUS MODELS

2.1 Review of the Noether Theorem

In the early part of this century, Emmy Noether (1918) discovered the fundamental invariance principle now known as the *Noether Theorem*. Noether not only derived the conservation (law) of total energy from a viewpoint of invariance, but also provided the formal methodology to study the general 'invariance' problem of a dynamic system. Influenced by the work of Klein (1918) and Lie (1891) on the transformation properties of differential equations invariant under continuous (Lie) groups, Noether had the ingenious insight of combining the methods of calculus of variations with those of Lie group theory. Since the first application of the Noether invariance principle to particle mechanics by Bessel-Hagan (1921), this area of mathematics has exhibited remarkable development in the last fifty years.

Let us consider a Lagrange function L, which is twice continuously differentiable in each of its $2n + 1$ arguments. We have the variational integral

$$J(x) = \int_a^b L(t, x(t), \dot{x}(t))\, dt, \tag{1}$$

where $x =$ the set of all vector functions $x(t) = (x^1 (t), \ldots, x^n (t))$, $t \in (a, b)$. Also consider the transformations (often Lie group transformations) given by

$$\bar{t} = \phi (t, x, \varepsilon), \quad \varepsilon = (\varepsilon^1, \ldots, \varepsilon^r),$$

$$\bar{x}^i = \psi^i (t, \chi, \varepsilon), \quad i = 1, \ldots, n \tag{2}$$

where $\varepsilon =$ the vector of r real, independent essential parameters. The "infinitesimal" transformations of (2) are obtained by expanding the right-hand side of (2) in a Taylor series around $\varepsilon = 0$ as

$$\bar{t} = t + \tau_s (t, x)\, \varepsilon^s + 0^1(\varepsilon)$$

$$\bar{x}^i = x^i + \xi_s^i (t, x)\, \varepsilon^s + 0 (\varepsilon) \quad i = 1, \ldots, n \tag{3}$$

$s = 1, \ldots, r$, summation convention in force.

Using the customary symbol of the infinitesimal transformations, we write (3) as

$$\chi_s = \tau_s (t, x)\, \frac{\partial}{\partial t} + \xi_s^i (t, x)\, \frac{\partial}{\partial x^i} + \left(\frac{d \xi_s^i}{d t} - \dot{x}^i\, \frac{d \tau_s}{d t} \right) \frac{\partial}{\partial x^i} . \tag{4}$$

The fundamental integral (1) in invariant under the r-parameter family of transformations (2) up to a divergence term, if there exist r functions Φ_s such that

$$
\begin{aligned}
L &= \left(\bar{t}, \bar{x}(\bar{t}), \frac{d\bar{x}(\bar{t})}{dt} \right) \frac{dt}{dt} - L(t, x(t), \dot{x}(t)) \\[2mm]
&= \varepsilon^s \frac{d\Phi_s}{dt} (t, x(t)) + 0 \, (\varepsilon) \, .
\end{aligned}
\tag{5}
$$

The *fundamental invariance identities* are given by

$$
\chi_s L + L \frac{d\tau_s}{dt} = \frac{d\Phi_s}{dt} \, ,
\tag{6a}
$$

or

$$
\frac{\partial L}{\partial t} \tau_s + \frac{\partial L}{\partial x^i} \xi_s^i + \frac{\partial L}{\partial \dot{x}^i} \left(\frac{d\xi_s^i}{dt} - \dot{x}^i \frac{d\tau_s}{dt} \right) + L \frac{d\tau_s}{dt} = \frac{d\Phi_s}{dt} \, .
\tag{6b}
$$

Noether's Theorem on Conservation Laws states that if the fundamental integral (1) of a problem in the calculus of variations is invariant up to a divergence term under the r-parameter family of transformations (2), then r distinct quantities Ω_s ($s = 1, \ldots, r$) are constant along any extremity and there exist r conservation laws,

$$
\Omega_s = -H \tau_s + \frac{\partial L}{\partial \dot{x}^i} \xi_s^i - \Phi_s = \text{const.} \ (s = 1, \ldots, r)
\tag{7}
$$

where H is the Hamiltonian defined by

$$
H = -L + \dot{x}^i \frac{\partial L}{\partial \dot{x}^i} \, .
\tag{8}
$$

The fundamental invariance identities given by (6) serve as the basic equation for determining the existence of the group of transformations under which the integral is invariant. Equation (6) is usually written, upon expanding its total derivatives, in the following system of partial differential equations in unknowns τ_s and ξ_s^i for any given Lagrangean function:

$$\frac{\partial L}{\partial t}\,\tau_s + \frac{\partial L}{\partial x^i}\,\xi_s^i + \frac{\partial L}{\partial x^i}\left[\frac{d\,\xi_s^i}{dt} + \frac{d\,\xi_s^i}{dx^j}\,\dot{x}^j - x^i\left(\frac{\partial\tau_s}{\partial\tau} + \dot{x}^j\frac{\partial\tau_s}{\partial x^j}\right)\right]$$

$$+ L\left(\frac{\partial\tau_s}{\partial t} + \frac{\partial\tau_s}{\partial x^j}\,\dot{x}^j\right) = \frac{d\Phi_s}{dt} \tag{9}$$

$$(s = 1, \ldots, r).$$

[See Sato (1981), pp. 242–251) for derivations of the Noether theorem and the invariance identities.]

2.2. Model 1: Zero Discount Rate

Frank Ramsey's original model, as well as the von Neumann model of capital accumulation developed by Samuelson (1970a, 1970b), and Liviatan-Samuelson's model of general neoclassical growth (1969) where the time variable t does not enter into the Lagrangean explicitly, are the typical examples belonging to this model. Model 1 can be represented by

$$\max \int_a^b L\,(x(t), \dot{x}(t))\,dt \tag{10}$$

where L satisfies all the necessary conditions for maximization.

The invariance identities for this problem are:

$$L\,\frac{d\tau_s}{dt} + \frac{\partial L}{\partial x^i}\,\xi_s^i + \frac{\partial L}{\partial \dot{x}^i}\left(\frac{d\xi_s^i}{dt} - \dot{x}^i\frac{d\tau_s}{dt}\right) = \frac{d\Phi_s}{dt}. \tag{11}$$

One of the solutions to the above give the well-known case of

$$\tau_i = \text{constant} = 1,$$
$$\xi_s^i \equiv 0,$$

(12)

which immediately leads to the conservation law of the Hamiltonian.

$$H = -L + \dot{x}^i \frac{\partial L}{\partial \dot{x}^i}$$

(13)

The Hamiltonian in the Ramsey model is the Ramsey rule for optimal saving. In the Liviatan-Samuelson model it is "a welfare measure of national income" in Kuznets' sense which should remain constant.

In a closed consumptionless system of the von Neumann type, L takes a special form

$$L\,(t, \lambda\,(t), K\,(t); \dot{\lambda}\,(t), \dot{K}\,(t)) = \dot{K}_1 + \lambda F\,[\dot{K}\,(t), K\,(t)]$$

(14)

where $K(t)$ = the vector of n capital goods $(K_1\,(t), \ldots, K_n\,(t))$, F = a smooth, neoclassical, concave, first-degree homogeneous transformation function (Samuelson, 1970a, 1970b), and λ = the multiplier. The Lie group transformations under which (14) is invariant are

$$\bar{t} = t + \varepsilon\tau\,(t, \lambda\,(t), K\,(t)),$$

$$\bar{\lambda}\,(t) = \lambda\,(t) + \varepsilon w\,(t, \lambda\,(t), K\,(t)),$$

(15)

$$\bar{K}_i\,(t) = K_i\,(t) + \varepsilon\,\xi^i\,(t, \lambda\,(t), K\,(t)). \qquad i = 1, \ldots, n$$

The invariance equation is

$$\frac{\partial L}{\partial \lambda}w + \frac{\partial L}{\partial K_i}\xi^i + \frac{\partial L}{\partial \dot{K}_i}\left(\frac{d\xi^i}{dt} - \dot{K}_i\frac{d\tau}{dt}\right) + L\,\frac{d\tau}{dt}$$

$$= \frac{d\Phi}{dt}(t, \lambda, K). \tag{16}$$

where

$$L = \dot{K}_1 + \lambda F,$$

$$\frac{\partial L}{\partial K_i} = \lambda\,\frac{\partial F}{\partial K_i},$$

$$\frac{\partial L}{\partial \dot{K}_1} = 1 + \lambda\,\frac{\partial F}{\partial \dot{K}_1},$$

and

$$\frac{\partial L}{\partial \dot{K}_j} = \lambda\,\frac{\partial F}{\partial \dot{K}_j}.$$

$$j = 2, ..., n.$$

Solving the above [see Sato (1981, pp. 279-285)], we get

$$\tau = \gamma,$$
$$\xi^i = \alpha K_i, \qquad i = 1, ..., n$$
$$w = -\alpha\lambda,$$
$$\Phi = \alpha K_1 + C. \qquad \alpha, \gamma = \text{constant}$$

Corresponding to the first parameter γ we have the conservation laws,

$$\gamma: \quad \Omega_1 = \lambda\,(F - \frac{\partial F}{\partial \dot{K}_1}\dot{K}_i) = \lambda(K_1\,\frac{\partial F}{\partial K_i}) = \lambda Y = \text{constant}, \tag{17}$$

$$i = 1, ..., n$$

where Y = national income, and to the second parameter a, we have

$$\alpha: \ \Omega_2 = - \lambda(\frac{\partial F}{\partial \dot{K}_i} K_i) = \lambda W = \text{constant} \tag{18}$$

where

$$W = \sum K_i P_i,$$

$$P_i = - \frac{\partial F}{\partial \dot{K}_i} = \text{supply price of } K_i.$$

Dividing Ω_1 by Ω_2,

$$\frac{\Omega_1}{\Omega_2} = \frac{Y}{W} \tag{19}$$

we get Samuelson's output-wealth conservation law [Samuelson (1970)].

2.3 Model 2: Fixed Discount Rate

The typical model of maximization of the discounted present value of welfare is represented by

$$\max \int_0^\infty e^{-\rho t} \ L\left(x\left(t\right), \dot{x}\left(t\right)\right) dt, \quad \rho > 0 \tag{20}$$

where L again satisfies all the conditions necessary for maximization. This model is most popular in economics and includes the extension of the Ramsey Model [see Cass (1965), Caton and Shell (1971), Liviatan and Samuelson (1969), and others], the neo-classical models of investment by Jorgenson (1967) and Lucas (1967) and the endogenous theory of technical change [see Sato and Suzawa (1983) and Sato and Ramachandran (1987)].

The general neo-classical optimal growth model takes the for

$$L(x(t), \dot{x}(t)) = U(c(t)) \tag{21}$$

and

$$c(t) = f(k(t), \dot{k}(t)) \tag{22}$$

where U = welfare function, c = consumption, k = capital per capita and f = the transformation function.

The general theory of investment takes the form

$$L(x(t), \dot{x}(t)) = pQ(L(t), K(t)) - WL(t) - \phi(\dot{K}(t), K) \tag{23}$$

where p = price of output Q, L = labor, K = capital, W = wage rate and ϕ = the adjustment function. The model of endogenous theory of technical change takes a similar form as (23); thus,

$$L(x(t), \dot{x}(t)) = P(Q)Q - C(Q, A) - \theta(Q, A, \dot{A}, B, \dot{B})$$
$$- \pi(Q, A, \dot{A}, B, \dot{B}) \tag{24}$$

where P = price of output Q, C = cost function, A = stock of applied knowledge, B = stock of basic knowledge (or research), θ = R&D expenditure for applied research and π = R&D expenditure for basic research.

One of the most important conservation laws hidden in the neo-classical growth model is the *income-wealth conservation law*, which was discovered by Weitzman (1976). This law was rediscovered by Samuelson (1982), Kemp and Long (1982), and Sato (1982). In Sato (1982), it was presented, for the first time, as a special case of the Noether theorem. We will follow the 1985 version of this approach [Sato (1985)].

By applying the infinitesimal transformation (3) on (20) when (20) takes the form (21), and by using the fundamental Noether invariance identities (6), we obtain

$$
e^{-\rho t}\left[-\rho U \tau + \frac{\partial U}{\partial k^i}\, \xi^i + \frac{\partial U}{\partial \dot{k}^i}\left(\frac{d\xi^i}{dt} - \dot{k}^i\, \frac{d\tau}{dt}\right) + U\, \frac{d\tau}{dt}\right] = \frac{d\Phi}{dt} \quad (25)
$$

Now assume that

$$
\begin{aligned}
\tau &= 1, \\
\xi^i &= 0. \qquad\qquad 1, \ldots, n
\end{aligned}
$$

Then we get, from (25) and from the Euler equation, along the optimal path,

$$
\frac{d\Phi}{dt} = -\rho\, e^{-\rho t}\, U \qquad (26)
$$

Also by differentiating (7) with t and setting $\tau = 1$ and $\xi^i = 0$, we have

$$
\frac{d\Phi}{dt} = \frac{d}{dt}\left[e^{-\rho t}\left(U - \dot{k}^i\, \frac{\partial U}{\partial \dot{k}^i}\right)\right] \qquad (27)
$$

Equating (26) and (27) and integrating both sides, we get

$$
e^{-\rho t}\left[U(t) - \dot{k}(t)\, \frac{\partial U(t)}{\partial \dot{k}(t)}\right] = \rho \int_t^\infty e^{-\rho s}\, U(s)\, ds \qquad (28)
$$

(The concavity and transversality conditions are implicitly assumed). By multiplying both sides of (28) by $e^{\rho t}$, we can derive the *income-wealth conservation law* as

$$
U - \dot{k}(t)\frac{\partial U(t)}{\partial \dot{k}(t)} = \rho \int_t^\infty e^{-\rho(s-t)}\, U(s)\, ds \qquad (29)
$$

or

$$\text{``Income''} = \rho \times \text{``wealth''} \qquad (30)$$

Alternately (29) can be rewritten [see Sato (1985, pp. 376-377)] as

$$e^{-\rho t}\left(U(t) - \dot{k}(t)\,\frac{\partial U(t)}{\partial \dot{k}(t)} \right) + \rho \int_0^t e^{-\rho s}\, U(s)\, ds = \rho \int_0^\infty e^{-\rho s}\, U(s)\, ds \qquad (31)$$

which can be interpreted as

"discounted income" + $\rho \times$ discounted stock of consumption
    = $\rho \times$ maximum discounted stock of consumption   (32)
    = constant.

We will leave it to the reader to come up with an appropriate economic interpretation of the conservation law that is inherent in the models of investment (4.4) and endogenous technical change (4.5).

2.4. Model 3: Variable Discount Rate

Samuelson (1982) pondered whether the income-wealth conservation law held when the discount rate varies with time. That is,

$$\text{``Income''} = \rho(t) \times \text{``wealth''} \qquad (33)$$

The Lie group approach enabled me to answer this question effectively [Sato (1985)]. Let the Lagrangean function be

$$L = e^{-\rho(t)}\, U(k(t), \dot{k}(t)) \qquad (34)$$

where $d\rho / dt$ is not necessarily constant. When $\rho_0(t) = \rho t$, $\rho_0 =$ constant, this model reduces to Model 2. Again by setting $\xi^i = 0$ and $\tau = 1$, we obtain [see Sato (1985, pp. 379-380)]

$$L - \dot{k}^i \, \frac{\partial L}{\partial \dot{k}^i} = \text{utility measure of generalized income}$$

$$= -\int_t^\infty \frac{\partial L}{\partial t} \, ds = \int_t^\infty \rho'(s) \, e^{-\rho(s)} \, U[k(s), \dot{k}(s)] \, ds$$

$$= \text{utility measure of generalized wealth.} \qquad (35)$$

It is apparent when $\rho'(s) = \rho = $ constant, we obtain Weitzman's standard income-wealth conservation law (30)

Now let

$$\tau = \frac{1}{\rho'(t)},$$
$$\qquad\qquad (36)$$
$$\xi^i = 0.$$

Then (35) becomes

$$\left[U - k^i \, \frac{\partial U}{k^i} \right] = \rho_t \int_t^\infty \left(\exp \left(-\int_t^s \rho_p \, dp \right) \left[U - \dot{k}^i \, \frac{\partial U}{\partial \dot{k}^i} \, \frac{d}{ds} \left(\frac{1}{\rho_s} \right) \right] \right) ds \quad (37)$$

or

$$\text{"income"} = \rho_t \times \text{"general wealth"} \qquad (38)$$

where

$$\rho_t = \frac{d\rho}{dt} \, \rho'(t).$$

The generalized wealth now includes capital gains and losses depending upon whether the last term of the integrated is positive or negative. This term depends on the supply price of investment $-\partial U / \partial k$, and the *variable* discount rate.

2.5 Model 4: Technical and Taste Change

Let the optimal control problem of the dynamic system be

$$\max \int_0^\infty D(t)\, L\,(x\,(t), \dot{x}\,(t), t)\, dt \tag{39}$$

Here L is directly affected by t, which represents taste and/or technical change. Note that there is an additional term $D(t)$. A special case of (39) is the usual optimal control problem of welfare maximization when L takes the form

$$D(t)\, L \; = \; D(t)\, U[c\,(k\,(t), \dot{k}\,(t), t)] \tag{40}$$

It is shown [Sato (1981, pp. 275-278)] that if $D(t)$ takes the usual form of $e^{-\rho t}$, then the "factor-augmenting" type of technical progress on k, i.e.,

$$C\,(k\,(t), \dot{k}\,(t), t) \; = \; c(k\,(\mathrm{t}), e^{\rho t}\, \dot{k}\,(t)) \tag{41}$$

implies the existence of the *conservation law for the current Hamiltonian*. That is,

$$\Omega = -\,(e^{-\rho t}\, U[c\,(k, e^{\rho t}\, \dot{k})] - U' \frac{\partial c}{\partial (e^{\rho t}\, \dot{k})}\, \dot{k})\; e^{\rho t} = \text{constant} \tag{42}$$

A more general case of

$$\max \int_0^\infty e^{-\rho t}\, U[k\,(t), \dot{k}\,(t), t]\, dt \tag{43}$$

where taste and technical change are not necessarily of the factor augmenting type, may be analyzed in the same manner as the previous model. The work by Sato, Nono and Mimura (1984) shows that when $\tau = e^{\rho t}$ and $\xi = b(t)\, \exp[\,-\int_0^t (U_k / U_{\dot{k}})\, ds]$, we have the conservation law:

$$U(t) - \dot{k}(t)\frac{\partial U}{\partial \dot{k}} + \int_t^\infty e^{-\rho(s-t)}\frac{\partial U}{\partial s}\, ds$$

$$= \rho \int_t^\infty e^{-\rho(s-t)}\, U(s)\, ds \tag{44}$$

which may be interpreted as

Income + "Value of Taste (Technical) Change" = $\rho \times$ wealth   (45)

There are many "hidden" conservation laws associated with the infinitesimal transformation related to the quantity variable $x(t)$. The reader may refer to Sato, Nono and Mimura (1984) for a special case when U does not explicitly contain t and

$$\tau = 0,$$
$$\xi \neq 0. \tag{46}$$

Here we have the *"modified" supply price conservation law.*

$$\Omega = \frac{\text{supply price of investment}}{\Theta'(k)} = \text{constant} \tag{47}$$

where Θ' is determined by $\xi \neq 0$ [see Sato, Nono and Mimura (1984, p. 40)].

2.6 Model 5: "Local" Conservation Laws

Most of the main results in economic dynamics are of *local nature,* we now look for what may be called the *"local" conservation laws* operating in the neighborhood of the stationary point. We assume that existence of the stationary equilibrium point corresponding to $\dot{k} = 0$, i.e., $k = k^*, \dot{k} = 0$. We begin with the local Lagrangean near $(0, k(0))$ in the form suggested by Samuelson (1972, p. 113, equation 46),

$$L = e^{-\rho t} (-1/2\, \dot{x}^2 - ax\dot{x} - 1/2x^2) \tag{48}$$

where $x = k(t) - k^*, \dot{x} = \dot{k}(t)$. The conservation law derived for the case [see Sato (1981)],

$$\tau = \delta = \text{constant}$$

$$\xi = \frac{\rho\delta}{2}x \tag{49}$$

is

$$\Omega = \frac{e^{-\rho t}}{2} \ [(\dot{x}^2 - x^2) - \rho x\,(\dot{x} + ax)], \quad \delta = 1$$

$$= H - \frac{\rho x}{2}\ \frac{\partial L}{\partial \dot{x}} = \text{constant}. \tag{50}$$

The value of the modified Hamiltonian (or income) remains constant, where the modification factor is equal to ρ *multiplied by the value of capital.* This is the local version of the income-wealth conservation law discussed in the earlier sections.

3. DISCRETE MODELS

3.1 Introduction

In applied mathematics and engineering, attention has also been paid to the analysis of discrete mechanics to determine whether or not continuous Hamiltonian and Lagrangean formalisms will apply and also whether or not the Noether theorem can be extended to the study of discrete mechanics (Maeda, 1982).

A few remarks are in order. First, in the study of discrete mechanics, the Noether theorem is not of great use because no infinitesimal change of time can be admitted [see Maeda (1982)], and because no energy-like integral can be derived from this approach. Hence, another method must be devised which will enable us to proceed. Secondly, as long as we focus our attention on the local aspects of the discrete dynamic system, we need not rely on the Noether-like theorem.

Section 2 presents a mathematical methodology to attack this problem. While no method is universally applicable, this method has proved to be very powerful, provided that one looks only for "local" conservation laws. We begin with a general description of the discrete-time growth model.

3.2 Model 6: Discrete Growth Models

We first present a general form of discrete-time N-sector growth models. Let t denote a discrete time variable $t = 0, 1, 2, \ldots$, and let $q = (q^1, q^2, \ldots, q^n)$ be a set of N capital goods measured in terms of unit of labor input. The symbol q_t means the value of q at the t th period. Given a well-defined social welfare (utility) index function, the society's objective is to maximize the sum of discounted utility over the infinite time horizon from $t = 0$ to $t = +\infty$. For simplicity we assume that no technical change is explicitly considered so that the welfare function does not contain t explicitly.

We begin by recalling a simple Ramsey model where the transformation function between consumption and capital accumulation is linear. Let f and U represent a neo-classical production function and a utility function respectively. Then, the society's welfare in the discrete model is represented by

$$U_t = U(q_t, q_{t+1}) = U[f(q_t) + q_t - (1+n)q_{t+1}] \qquad (51)$$

where n is the growth rate of labor (Samuelson, 1967). The society maximizes the discounted sum of welfare "functional"

$$\sum_{t=0}^{\infty} (1 + \delta)^{-t} U_t = \sum_{t=0}^{\infty} \lambda^{-t} U_t, \quad \lambda = 1 + \delta \qquad (52)$$

where $\delta \geq 0$ is a fixed time discount rate. As is usual, it is assumed that f and U satisfy $f > 0, f' > 0, f'' < 0; U' > 0, U'' < 0$. Also it is implicitly assumed that q, a set of capital goods, takes a non-negative value satisfying certain inequality conditions. The optimal growth path generated by the maximization of (52) is assumed to have an equilibrium solution q^*, which enables one to look for conservation laws operating near the equilibrium point. By adopting a new variable $q_t - q^*$, rather than q itself, one can look for "local" conservation laws operating near the equilibrium point q^*. By approximating the original utility function up to the second order, one can study the dynamic behavior with the approximated Lagrangean function which is quadratic homogeneous in $q_t - q^*$. As long as we deal with a local area, no inequality condition on q_t is necessary.

As there is no need to confine the analysis to the Ramsey-type model, we write a general discrete version of the Samuelson-Solow model and define the society's objective welfare "functional" as

$$\max \sum_{t=0}^{\infty} \lambda^{-t} L(q_t, q_{t+1}), \quad \lambda = 1 + \delta \qquad (53)$$

where L denotes a social welfare function of C^2 class satisfying the appropriate conditions. Assume also that the variable (q_t, q_{t+1}) is contained in an admissible convex set and the equilibrium point q^* is admitted in the domain. The function L is assumed to be strictly concave and its Hessian matrix at the equilibrium point is negative definite.

Let v_t denote the forward difference of q_t, i.e., $v_t = q_{t+1} - q_t$. For notational convenience, we let q_t stand for $q_t - q^*$ and v_t stand for $q_{t+1} - q_t$.

Also, in order to formulate the system in terms of discrete Lagrangean mechanics, we consider L as a function of (q_t, v_t). By approximating the Lagrangean near the equilibrium point up to the second order, the problem of maximizing the society's welfare function is reduced to

$$\max \sum \lambda^{-t} L_t(q_t, v_t), \quad L_t = \frac{v' B_0 v}{2} + v' C_0 q + \frac{q' D_0 q}{2} \tag{54}$$

where B_0 and D_0 are symmetric, and B_0 and $B_0 - C_0$ are nonsingular matrices. The elements of B_0, C_0 and D_0 are the cross-derivatives of the original function in (53) evaluated at (q^*, q^*) as

$$b_{ij} = \frac{\partial^2 L}{\partial q^i_{t+1} \partial q^j_{t+1}} (q^*, q^*),$$

$$c_{ij} = b_{ij} + \frac{\partial^2 L}{\partial q^i_{t+1} \partial q^j_t} (q^*, q^*),$$

$$d_{ij} = c_{ij} + \frac{\partial^2 L}{\partial q^j_{t+1} \partial q^i_t} (q^*, q^*) + \frac{\partial^2 L}{\partial q^i_t \partial q^j_t} (q^*, q^*) \tag{55}$$

The next task is to make the discounted social welfare function $\lambda^{-t}L$ *independent of t*. We introduce a new variable Q in place of q as

$$Q_t = \lambda^{-t/2} q_t \qquad (56)$$

and also V_t, the forward difference of Q_t as

$$V_t = Q_{t+1} - Q_t = \lambda^{-t/2} [\lambda^{-1/2} v_t + (\lambda^{-1/2} - 1) q_t]. \qquad (57)$$

Then the welfare function becomes quadratic homogeneous in Q and V:

$$\lambda^{-t} L = \tilde{L} = \frac{V' BV}{2} + V' CQ + \frac{Q' DQ}{2} \qquad (58)$$

where

$$B = \lambda B_0,$$
$$C = (\lambda - \lambda^{1/2}) B_0 + \lambda^{1/2} C_0 \qquad (59)$$
$$D = (\lambda^{1/2} - 1)^2 B_0 + (\lambda^{1/2} - 1) (C_0 + C_0') + D_0$$

The discrete Euler equation (Euler difference equation) for maximization is transformed to a Hamiltonian system by the Legendre transformation (Maeda, 1980). If we introduce the implicit prices P or "momentums," conjugate to V (not v), as

$$P_{t+1} = \frac{\partial \tilde{L}_t}{\partial V} = BV_t + CQ_t \qquad (60)$$

then, the Euler difference equation may be written as a linear system in (Q, P).

We are now in a position to study the dynamic behavior of the model near the equilibrium point. After some tedious calculations, we obtain the linearized equation in the Hamiltonian formalism as

$$x_{t+1} = Ax_t \tag{61a}$$

where

$$A = \begin{bmatrix} -(B - C')^{-1} & I \\ -B\,(B - C')^{-1} & C \end{bmatrix} \begin{bmatrix} C - D & -I \\ I & 0 \end{bmatrix} \tag{61b}$$

and

$$x' = (Q^1, \dots, Q^n, P_1, \dots, P_n) \tag{61c}$$

According to the theory of discrete dynamics (Maeda, 1980 and 1982), the transition matrix A is a *symplectic matrix;* that is, it satisfies

$$A' J A = J, \quad \text{where } J = \begin{bmatrix} 0 & I \\ -I & 0 \end{bmatrix} \tag{62}$$

In order to derive conservation laws associated with the linear dynamic system given by (61), we must construct first integrals of the form

$$f_s(x) = \frac{x'Sx}{2}, \quad S' = S \tag{63}$$

such that it satisfies the condition

$$f_s(x_{t+1}) = f_s(x_t), \quad \text{for all } t. \tag{64}$$

In what follows, we digress to show how such integrals can be obtained.

3.3 Quadratic Conservatives: A Mathematical Digression[1]

The problem of optimizing the functional;

$$\sum_{t=0}^{T} L_t(q_t, q_{t+1}, t) = \sum_{t=0}^{T} L_t(q_t, v_t, t) \tag{65}$$

through the discrete variational principle will yield the system of Euler difference equations:

$$\frac{\partial L_t}{\partial v_t^i} - \frac{\partial L_{t-1}}{\partial v_t^i} - \frac{\partial L_t}{\partial q_t^i} = 0 \tag{66}$$

where $v_t^i = q_{t+1}^i - q_t^i$ is the forward difference of the coordinates q_t^i. Like the continuous case there are two formalisms: the Lagrangean and Hamiltonian formalisms. Any system in one formalism can be expressed in the other formalism under the appropriate conditions. Suppose that a Lagrangean function satisfies

$$\det\left(\frac{\partial^2 L}{\partial v^i \partial v^j}\right) \neq 0, \quad \det\left(\frac{\partial^2 L}{\partial v^i \partial v^j} - \frac{\partial^2 L}{\partial v^i \partial q^j}\right) \neq 0.$$

Then by putting

$$p_{i,t+1} = \frac{\partial L_t}{\partial v_t^i} \tag{67a}$$

and

$$H(q_t, p_{t+1}) = p_{i,t+1} v_t^i - L(q_t, v_t) \tag{67b}$$

(The summation convention in force.)

equation (66) can be transformed to

$$q_{t+1}^i - q_t^i = \frac{\partial H}{\partial p_{it}}, \quad p_{i,t+1} - p_{it} = -\frac{\partial H}{\partial q_t^i} \tag{68}$$

By a "discrete system" it is meant a system described by ordinary difference equations of the first order. Equation (68) is a discrete system generated from the optimization. In general the discrete system is expressed as $x_{t+1} = \phi(x_t)$. A point sequence $\{x_t\}$ subject to the discrete system is called a solution, and a real smooth function is called a first integral (F.I.) if its value remains constant along any solution. The discrete system has two important properties intrinsic to any Hamiltonian system (Abraham and Marsden, 1978):

1. If f and g are F.I.'s, so is $\{f, g\}$, where $\{,\}$ denotes the Poisson bracket;

2. If h is an F.I., then X_h is a symmetry operator where X_h is the Hamiltonian operator with the Hamiltonian h.

Here, a symmetry operator of a discrete system is defined to be an infinitesimal operator such that its flow commutes with the mapping corresponding to the discrete system. It is shown that a set of all F.I.'s and all symmetry operators form Lie algebras respectively, and that the mapping $f \to -X_f$ gives a Lie algebra homomorphism (Maeda, 1980).

A discrete linear Hamiltonian system is called symplectic, as it preserves the standard symplectic structure $\omega = dp_i \wedge dq^i$ (Sato, 1981). The discrete system corresponding to a linear symplectic mapping is expressed in the normal form as

$$x_{t+1} = A \cdot x_t \tag{69}$$

where $x' = (q^1, q^2, ..., q^n, p_1, p_2, ..., p_n)$.

Let G and g denote the 2N-dimensional symplectic group $Sp(N, \mathbf{R})$ and its Lie algebra, respectively:

$$G = \{A \in M(2N, \mathbf{R}) \mid A'JA = J\},$$
$$g = \{X \in M(2N, \mathbf{R}) \mid X'J + JX = 0\},$$

where $J = \begin{bmatrix} 0 & -I \\ I & 0 \end{bmatrix}$

and dash denotes matrix transpose. Our problem is to establish an algebraic approach to finding the quadratic form

$$f_s(x) = \frac{x'Sx}{2}, \qquad S' = S, \tag{70}$$

conserved along any solution of the linear recurrence on $\mathbf{R}^N$

$$x_{t+1} = Ax_t, \tag{71}$$

where A is an arbitrary element of G. The whole of conservatives given by (70) forms a Lie algebra with respect to the Poisson bracket (Maeda, 1980).

Linear space Ξ. We introduce a linear space Ξ of all matrices commuting with A, and it is proved that one of its subspaces is Lie algebra isomorphic to the whole of quadratic conservatives given by (70).

Now, (70) is conserved along any solution of (71), if and only if A and JS commute. Then, the whole of quadratic conservatives is identified with the following linear space of all coefficient matrices:

$$\Omega = \{S \in M(2N, \mathbf{R}) \mid [A, JS] = 0, \ S' = S]\},$$

where $[A, B] = AB - BA$. Ω forms a Lie algebra with respect to the bracket

$$< S, T > = SJT - TJS, \tag{72}$$

which is a respresentation of the Poisson bracket on Ω. Apart from looking into Ω directly, we introduce a linear space Ξ of all matrices that commute with A :

$$\Xi = \{L \in M(2N, \mathbf{R}) \mid [A, L] = 0\}.$$

We define two linear mappings $\eta\colon \Xi \to \Xi$ and $\sigma\colon \Xi \to \Omega$ by

$$\eta(L) = JL'J, \quad \sigma(L) = J(L + \eta(L))\,/2 = (JL + (JL)')\,/2. \tag{73}$$

Lemma 1. $\eta^2 = id., \ \eta(\Xi) = \Xi.$

Proof. Let $L \in \Xi$. Then, it follows from direct calculation that $\eta^2(L)$ $= L$ and $[\eta(A), \eta(L)] = \eta([A, L]) = 0$. Since A is symplectic, we have $\eta(A) = -A^{-1}$ and accordingly $[A, \eta(L)] = 0$, which means $\eta(\Xi) \subset \Xi$. This together with $\eta^2 = id.$ leads to $\eta(\Xi) = \Xi$.

The lemma shows that η is an involution map on Ξ. Then, η has two eigenvalues ± 1 and Ξ is a direct sum of the two. That is, $\Xi = \Theta \,\dot{+}\, \Phi$, where

$$\Theta = \{L \in \Xi \mid \eta(L) = L\}, \quad \Phi = \{L \in \Xi \mid \eta(L) = -L\}.$$

The projector P from Ξ onto Θ is given by

$$P(L) = (L + \eta(L))\,/2$$

The condition $\eta(L) = L$ is equivalent to $L'J + JL = 0$ so that Θ, which is an intersection of g and Ξ, is a subalgebra of $sp(N, \mathbf{R})$. Then P produces an element of g from any matrix commuting with A. It is to be stressed that A is an element of a Lie group and $P(L)$ is an element of a Lie algebra.

Now, Ξ is connected with Ω in the following manner:

Lemma 2. $\sigma(\Xi) = \Omega, \ \sigma^{-1}(0) = \Phi.$

Proof. We choose an arbitrary L. Be definition, $\sigma(L)$ is symmetric. Moreover, since $J\sigma(L) = -(L + \eta(L))\,/2$, it belongs to Ξ and commutes with A. Thus $\sigma(L) \in \Omega$. Conversely, for any $S \in \Omega$, we put $L = -JS$. Then, it is easily proved that $L \in \Xi$ and $\sigma(L) = S$. The second assertion is obvious from the definitions of σ and Φ.

Since $\Xi = \Theta \dotplus \Phi$, Lemma 2 shows that Ω is linearly isomorphic to Θ. Furthermore, it holds after slight calculation that

$$\sigma([L, M]) = \langle \sigma(L), \sigma(M) \rangle, \tag{74}$$

where $<, >$ is given by (72), and L and M belong to Θ. Combining this and Lemma 2, we have

Lemma 3. *Θ is Lie algebra isomorphic to Ω.*

The linear mapping σ restricted on Θ gives a momentum mapping $\hat{J}$ in symmetry reduction theory of classical mechanics (Abraham and Marsden, 1978). Furthermore, we note that

$$\sigma(P(L)) = \sigma(L) \tag{75}$$

holds for any $L \in \Xi$.

A subspace Ξ_1 of Ξ. We now study how many invariants are obtained among polynomials in A. To see this, we define a linear subspace Ξ_1 of Ξ.

$$\Xi_1 = \mathrm{span}\ \{I, A, A^{\pm 1}, A^{\pm 2}, \dots\}.$$

Hereafter, we denote by ϕ_A and f_A the minimal polynomial and the eigenpolynomial of A, respectively. When ϕ_A is equal to f_A, any matrix that commutes with A is expressed as a polynomial in A, so that Ξ_1 coincides with Ξ. Almost any element in G has this property.

Now, we put

$$\Theta_1 = \Xi_1 \cap \Theta, \qquad \Phi_1 = \Xi_1 \cap \Phi, \qquad \Omega_1 = \sigma(\Xi_1) = \sigma(\Theta_1).$$

Our interest centers in dim Ω_1 ($= \dim \Theta_1$), which is the number of linearly independent conservatives (63) obtained from polynomials in A.

Lemma 4. *Let* $d = dim \; \Omega$, *and* $k = deg \; \phi_A$, *and one of the following three cases holds good:*

1. If $k = 2s + 1$, *then* $d = s$.

2. If $k = 2s$ *and* $\phi_A(0) = 1$, *then* $d = s$.

3. If $k = 2s$ *and* $\phi_A(0) = -1$, *then* $d = s-1$.

Here, s is an integer.

Proof. We note that k is equal to dim Ξ_1. For any integer i, we put

$$B_i = A^i - A^{-i} \tag{76}$$

and $C_i = A^i + A^{-i}$. Then, it holds that $B_i \in \Theta_1$ and $C_i \in \Phi_1$. When k is an odd number $2s + 1$, $\{B_1, ..., B_s, C_0, ..., C_s\}$ forms a basis of Ξ_1, and we have $k = s$. Next, when k is equal to $2s$, the $2s - 1$ matrices $\{B_1, ..., B_{s-1}, C_0, ..., C_{s-1}\}$ are linearly independent and further either B_s or C_s is linearly independent of these. Now, since A is symplectic, its minimal polynomial ϕ_A satisfies $\phi_A(A) = \phi_A(A^{-1}) = 0$. Then ϕ_A must take one of the following two forms:

(a) $(x^{2s} + 1) + a_1(x^{2s-1} + x) + ... + a_{s-1}(x^{s+1} + x^{s-1}) + a_s x^s$,

(b) $(x^{2s} - 1) + a_1(x^{2s-1} - x) + ... + a_{s-1}(x^{s+1} - x^{s-1})$.

When $\phi_A(0) = 1$ and (a) holds, B_s becomes linearly independent. When $\phi_A(0) = -1$ and (b) holds, C_s does.

Next, we propose a simple scheme to construct a basis of Ω_1. If we do not know dim Θ_1 in advance, this scheme naturally produces a maximum number of linearly independent elements.

Lemma 5. *Suppose that* $\{\sigma(A), \; \sigma(A^2), \; ..., \; \sigma(A^d)\}$ *are linearly independent and* $\sigma(A^{d+1})$ *are linearly dependent on these d matrices. Then, the former d matrices form a basis of* Ω_1.

Proof. Since A^i belongs to G, we have $\sigma(A^i) = JB_i / 2$, where B_i is given by (76). As is seen in the proof of Lemma 4, when dim $\Theta_1 = d$, the set $\{B_1, ..., B_d\}$ forms a basis of Θ_1, and the converse is true. Since $(1/2) J \therefore \theta_1 \rightarrow \Theta_1$ gives a linear isomorphism, the assertion is verified.

Again, we remark that for almost every element A of G, its minimal polynomial ϕ_A coincides with the eigenpolynomial f_A. In this case, any matrix commuting with A is expressed as a polynomial in A so that Ω_1 is Ω itself.

Lemma 6. *Suppose that for an element A of G, its minimal polynomial coincides with its eigenpolynomial. Then, $\{\sigma(A), \ldots, \sigma(A^N)\}$ forms a basis of Ω.*

Proof. Under the supposition, it holds that dim $\Omega_1 = 2N$ and $\phi_A(0)$ $= f_A(0) = \det(A) = 1$. Then, we have the conclusion from Lemma 4.

We can obtain all quadratic conservatives for $A \in G$ in the case of this theorem. Furthermore, we have

Lemma 7. *Under the same condition as in Lemma 6, the linear discrete system (69) is completely integrable (Maeda, 1987).*

Proof. Since $P(A^i) = (A^i - A^{-i}) / 2$, we have $[P(A^i), P(A^j)] = 0$. Then, it follows from (74) and (75) that $\langle \sigma(A^i), \sigma(A^j) \rangle = 0$. That is, the system (69) admits N mutually commutative conservatives.

We have seen that the generic linear Hamiltonian system admits as many quadratic F.I.'s as the degree of freedom. If two of the eigenvalues of the coefficient matrix A coincide, another kind of quadratic F.I.'s may be obtained. We do not, however, venture into this type of problems in this chapter.

Example. It may be of some use to illustrate the methodology developed in this section by a simple example. Consider a linear Hamiltonian system with one degree of freedom.

$$q_{t+1} = aq_t + bp_t,$$

$$P_{t+1} = cq_t + dp_t,$$

where a, b, c and d are real constants subject to $ad - bc = 1$. In this case

$$A = \begin{bmatrix} a & b \\ c & d \end{bmatrix} \in Sp(1, P).$$

The hypothesis of Lemma 6, if and only if $a + d \neq \pm 2$.

Hence, it follows that $\Omega = \Omega_1$, and that its dimension is one. Furthermore we can calculate $\sigma(A)$ as

$$\sigma(A) = \begin{bmatrix} c & \dfrac{(d-a)}{2} \\ \dfrac{(d-a)}{2} & -b \end{bmatrix}$$

Then the corresponding F.I. is given by[2]

$$f_{\sigma(A)} = 1/2[cq^2 + (d - a)\, qp - bq^2].$$

We say also that Lemma 6 assures that the system in this example admits one and only one linearly independent quadratic F.I. We note that when $d \neq 0$, the system is expressed in the form (68) as

$$q_{t+1} - q_t = \frac{\partial H}{\partial p}\, (q_t, p_{t+1}), \quad p_{t+1} - p_t = -\frac{\partial H}{\partial q}(q_t, p_{t+1})$$

where H is given by the following quadratic form.

$$H = \frac{b}{2d}\, p^2 + \frac{1-d}{d}\, qp - \frac{c}{2d}\, q^2.$$

4. ECONOMIC CONSERVATION LAWS

Let us recall the linear Hamiltonian system (61). Our aim is to derive $f_{\sigma(A)}$ and to represent it in terms of the original variables (q, v), which is the conservation law in the system.

The matrix A in (61b) belongs to $S_p(N, \mathbf{R})$. Therefore, we can apply Lemma 6 to the system to obtain quadratic F.I.'s in terms of (Q, P) by calculating the coefficient matrices $\sigma(A^j)$ ($j = 1, 2, \ldots$). The F.I.'s thus obtained will be rewritten as function of (q_t, q_{t+1}) or (q_t, v_t) through the coordinate transformation given by (56), (57) and (60). Combining (56), (57) and (60), we obtain the transformation relating (Q, P) with (q, v) or with (q_t, q_{t+1}) as

$$\begin{bmatrix} Q \\ P \end{bmatrix} = \lambda^{-t/2} T \begin{bmatrix} q \\ v \end{bmatrix} \tag{77a}$$

where

$$T = \begin{bmatrix} I & 0 \\ (C-D) + (\lambda^{-1/2} - 1)(B-C') & \lambda^{-1/2}(B - C') \end{bmatrix} \tag{77b}$$

or

$$\begin{bmatrix} Q_t \\ P_t \end{bmatrix} = \lambda^{-t/2} \tilde{T} \begin{bmatrix} q_t \\ \lambda^{-t/2} q_{t+1} \end{bmatrix} \tag{78a}$$

where

$$\tilde{T} = \begin{bmatrix} I & 0 \\ (C-D) - (B-C') & (B - C') \end{bmatrix} \tag{78b}$$

Then calculate $T'JAT$ and express its elements in terms of B_0, C_0 and D_0 to obtain

$$T'JAT = \lambda^{-1/2} \begin{bmatrix} \lambda C_0 - C'_0 + D_0 & \lambda B_0 - C'_0 + D_0 \\ -B_0 + C_0 & -B_0 + C_0 \end{bmatrix} \quad (79)$$

where $J = \begin{bmatrix} 0 & I \\ -I & 0 \end{bmatrix}$ as defined by equation (62).

Recall that $\sigma(A) = [JA + (JA)'] / 2$ and then we have

$$T'\sigma(A)T = \frac{T'JAT + (T'JAT)'}{2} \quad (80)$$

By setting $\lambda f_{\sigma(A)} = I_1$, we obtain the following quadratic polynomial I_1, which is the conservation law in the system:

$$I_1 = \lambda^{-t} \left[\frac{v'(-B_0 + C_0)v}{2} + \frac{v'((\lambda-1)B_0 + D_0)q}{2} + \frac{q'(\lambda-1)C_0 + D_0)q}{2} \right] (81)$$

This conservation law can be expressed in a simple form using the original Lagrangean (54) as

$$I_1 = \lambda^{-t} \left[-(v'\frac{\partial L}{\partial v} - L) + \frac{1}{2} v' \frac{\partial L}{\partial q} + \frac{(\lambda-1)}{2} q' \frac{\partial L}{\partial v} \right] \quad (82)$$

Alternatively using (q_t, q_{t+1}) and the transformation T, we can derive the conservation law from

$$\tilde{T}JAT = \begin{bmatrix} -(B - C) & 2B - C + D - C' \\ 0 & -(B - C) \end{bmatrix}. \quad (83)$$

Due to the definition of B, C and D, each component of minor matrices in the above is given by

$$-(B - C)_{ij} = \lambda^{1/2} \alpha_{ij}, \qquad (2B - C + D - C')_{ij} = \beta_{ij}$$

where α_{ij} and β_{ij} are calculated from the cross-derivatives of $L(q_t, q_{t+1})$ in (53):

$$\alpha_{ij} = \frac{\partial^2 L}{\partial q_{t+1}^i \partial q_t^i} (q^*, q^*),$$

$$\beta_{ij} = \lambda \frac{\partial^2 L}{\partial q_{t+1}^i \partial q_{t+1}^j} (q^*, q^*) + \frac{\partial^2 L}{\partial q_t^i \partial q_t^j} (q^*, q^*).$$

Moreover, since $\sigma(A)$ is the symmetrization of JA, $y'\sigma(A)y = y'JAy$ holds for any 2N-tripled column vector y. Then from (78) and (83), we obtain the conservation law in terms of (q_t, q_{t+1}) as

$$\tilde{I}_1 = \lambda^{-1}[\lambda \alpha_{ij} q_t^i q_t^j + \beta_{ij} q_t^i q_{t+1}^j + \alpha_{ij} q_{t+1}^i q_{t+1}^j] \qquad (84)$$

(The summation convention in force.)

The concavity of the Lagrangean assures that $\tilde{I}_1$ can be expressed in terms of (q_t, v_t) as I_1 in equation (81).

Theorem. *Every discrete-time optimal growth Model given by (53) admits a local conservation law (82).*

The economic meaning of (82) is very similar to the conservation law for the continuous-time model [Sato (1981, p. 264)]. It is to be noted, however, that the term $1/2v'(\partial L / \partial q)$ is additional. This term

$$\sum \frac{(\text{capital accumulation}) \times (\text{marginal utility of consumption})}{2}$$

= sum of the values of investment measured in terms of welfare,

is considered as a correction term against the finiteness of the time period. It should also be noted that when $\lambda = 1$ (no discount rate), the conservation law is reduced to

$$I_0 = -(v' \frac{\partial L}{\partial v} - L) + \frac{1}{2} v' \frac{\partial L}{\partial q} \tag{85}$$

which is expressed in terms of q_t and q_{t+1} as

$$I_0 = -\frac{1}{2} (q_{t+1} - q_t)' \left(\frac{\partial}{\partial q_{t+1}} - \frac{\partial}{\partial q_t} \right) L_t - L_t \tag{86}$$

When $\lambda \neq 1$ (more specifically $\lambda > 1$), the third term in (82) measures the modifying factor due to the discount rate, very similar to the continuous case [Sato (1981, p. 264)].

Finally, let us work out in detail the conservation law associated with the discrete Ramsey model defined by (51). The model is a one-sector model and is assured to have one linearly independent quadratic F.I. (or the conservation law).

In this case, the coefficients α and β in (84) are given by

$$\alpha = \frac{\partial^2 L}{\partial q_{t+1} \partial q_t} (q^*, q^*) = -(1 + n) U''(c^*) (1 + f'(q^*))$$

$$\beta = \lambda \frac{\partial^2 L}{\partial q_{t+1}^2} (q^*, q^*) + \frac{\partial^2 L}{\partial q_t^2} (q^*, q^*)$$

$$= \{ \lambda (1 + n)^2 + (1 + f'(q^*))^2 \} U''(c^*) + f''c(q^*) U'(c^*)$$

where $c^* = f(q^*) - nq^*$. The Euler difference equation is reduced to

$$\alpha q_{t+1} + \beta q_t + \lambda \alpha q_{t-1} = 0.$$

The linearized discrete canonical equations are given by

$$
\begin{bmatrix} q_{t+1} \\ p_{t+1} \end{bmatrix} =
\begin{bmatrix} \dfrac{\xi - \eta}{\xi} - \dfrac{\eta^2 - \xi\zeta}{\xi(\xi - \eta)} & \dfrac{\lambda t}{\xi - \eta} \\ -\lambda^t \times \dfrac{\eta^2 - \xi\zeta}{\xi - \eta} = & \dfrac{\xi}{\xi - v} \end{bmatrix}
\begin{bmatrix} q_t \\ p_t \end{bmatrix}
$$

where

$$\xi = \frac{\partial^2 L_t}{\partial q_{t+1}^2}(q^*, q^*)$$

$$\eta = \frac{\partial^2 L_t}{\partial q_{t+1}^2}(q^*, q^*) + \frac{\partial^2 L_t}{\partial q_{t+1}\, \partial q_t}(q^*, q^*)$$

$$\zeta = \frac{\partial^2 L_t}{\partial q_{t+1}^2}(q^*, q^*) + 2\frac{\partial^2 L_t}{\partial q_{t+1}\, \partial q_t}(q^*, q^*) + \frac{\partial^2 L_t}{\partial q_t^2}(q^*, q^*)$$

The expressions ξ, η and ζ are related with α and β by

$$\alpha = \eta - \xi, \qquad \beta = (1 + \lambda)\xi - 2\eta + \zeta.$$

The characteristic roots associated with the canonical equations can be derived from

$$\alpha x^2 - (2\alpha - \zeta)x + \alpha = 0.$$

The reader can show that under the usual assumptions on f and v, the two roots are positive with one root greater than unity and the other root less than unity—the saddle point property.[3]

The conservation law is now given by

$$I_1(q_t, q_{t+1}) = 1/2\,\lambda^{-t}(\alpha q_{t+1}^2 + \beta q_{t+1}q_t + \lambda\alpha q_t^2).$$

One can easily check the validity of this conservation law from the Euler difference equation. Since $S' \times \lambda^{-t} \times M \times S = \lambda^{-(t-1)}M$, we have

$$I_1(q_t, q_{t+1}) = I_1(q_{t-1}, q_t),$$

where

$$S = \begin{bmatrix} -\beta/\alpha & \lambda \\ 1 & 0 \end{bmatrix},$$

$$M = \begin{bmatrix} \alpha & \beta/2 \\ \beta/2 & \lambda\alpha \end{bmatrix}.$$

We close Part II by adding a few remarks on discrete systems. First, the discrete Noether theorem is not as powerful as in the continuous case. Second, the method presented here is not almighty either. It is a powerful and systematic method as long as one deals with linear systems. Finally, through the transformations (56) and (57), the system is reduced to the one independent of t explicitly. This was possible because the term involving t is multiplicative in the Lagrangean. If a technical change factor is introduced in the system in a general manner, then there may be no transformations which reduce the system to the desirable form.

5. SUMMARY

The purpose of this article is first to show how powerful the Noether theorem and Lie groups are in discovering both hidden and unhidden symmetries and conservation laws in continuous dynamic systems. Second, the discrete time optimization case requires a new methodology especially useful in linear systems. The resulting conclusions for the discrete systems are much more complicated than the continuous case. The main results of this paper are summarized in the following table.

Summary of Conservation Laws Table

	Lagrangean	Infinitesimal Transformation	Conservation Laws	Examples
Model I	$L = L\,(x(t), \dot{x}(t))$	$\tau = 1$ $\xi = 0$	H = Wealth Measure of National Income = constant	Original Ramsey Model
	$L = \dot{K}_1 + \lambda F(K, \dot{K})$	$\tau = \gamma$ = constant $\xi^i = \alpha K_i$ $w = -\alpha\lambda$ $\Phi = \alpha K_1 + C$ α, C = constant	λY = constant λW = constant i.e., $\dfrac{Y}{W} = \dfrac{\text{output}}{\text{wealth}}$ = constant	von Neumann- Samuelson Model
Model II	$e^{-\rho t} L(x(t), \dot{x}(t))$ $\rho > 0$	$\tau = 1$ $\xi = 0$ $\Phi \neq 0$ $\dfrac{d\Phi}{dt} = -\rho e^{-\rho t} L$	Income – Wealth Conservation Law = Discounted Income + ρ × Discounted Stock of Consumption = ρ × Max Discounted Stock of Consumption = constant	Neoclassical Growth Model (Weitzman) Neoclassical Theory of Investment Endogenous Theory of Technical Change
Model III	$e^{-\rho_t\,(t)} L(x, \dot{x})$	$\tau = \dfrac{1}{\rho'(t)}$ $\xi^i = 0$	Income = $\rho'(t)$ × "generalized" wealth	Variable Discount Rate (Samuelson, Sato)

Summary of Conservation Laws Table—continued

	Lagrangean	Infinitesimal Transformation	Conservation Laws	Examples
Model IV	$e^{-\rho t} L(x, x, t)$ $= e^{-\rho t} L(x, e^{\rho t} \dot{x})$	$\tau = e^{\rho t}$ $\xi = 0$ $= $ constant	Current Hamiltonian	"Factor-Aug. Technical Change on k" (Sato)
	General Case $e^{-\rho t} L(x, \dot{x}, t)$	$\tau = e^{\rho t}$ $\xi = b(t) \times$ $\exp \left(\int_0^t \left(-\dfrac{L_x}{L_{\dot{x}}} \, ds \right) \right)$	Income + "Value of Taste (Technical) Change" $= \rho \times$ wealth	General Technical and Taste Change (Sato, Nôno and Mimura)
		$\tau = 0$ $\xi \neq 0$	Modified Supply Price of Investment	*same as above*
Model V	$e^{-\rho t} [Q]$ $Q = $ Quadratic in x and $\dot{x}$	$\tau = \delta = $ constant $\xi = \dfrac{\rho \delta}{2} \xi$ $= $ constant	Modified Hamiltonian: Income + $\rho \times$ value of capital	Near the Steady-State (Sato)
Model VI	$\sum \lambda^{-t} L_t$ $L_t = $ Quadratic in $x(t) = q(t) - q^*$ and $x(t+1) - x(t) = v(t)$ $\lambda = 1 + \delta$		Modified Hamiltonian (1) Discrete Model Modification and (2) Discount Factor Modification $\lambda^{-t} [(-v' \dfrac{\partial L}{\partial v} - L)$ $+ 1/2 \, v' \dfrac{\partial L}{\partial q}$ $+ \dfrac{(\lambda^{-1})}{2} q' \dfrac{\partial L}{\partial q}$ $= $ constant	Discrete Dynamic System (Local Approximation) (Sato-Maeda)

NOTES

1. This section is largely due to Maeda (1988).
2. As J in this section (Section 3) is simply (-1) multiplied by J in equation (62), $\sigma(A)$ and $f_{\sigma(A)}$ should be multiplied by (-1) if J in equation (62) is to be used.
3. The reader can also check the saddlepoint property directly from the Euler difference equation using $1 + f'(q^*) = \lambda (1 + n)$.

REFERENCES

Abraham, R. and Marsden, J. E. (1978). *Foundations of Mechanics*, 2nd edition, Benjamin.

Bessel-Hagen, E. (1921). Über die erhaltungssätze der Elektrodynamik, *Math. Ann., 84*, 258-276.

Caton, C. and Shell, K. (1971). An exercise in the theory of heterogeneous capital accumulation, *Review of Economic Studies, 38*, 13-22.

Jorgenson, D. W. (1967). Theory of Investment Behavior, *Determinants of Investment Behavior*, R. Ferber (editor), New York: NBER.

Kemp, M. C. and Long, N. V. (1982). On the evaluation of social income in a dynamic economy. *Samuelson and Neoclassical Economics*. G. R. Feiwel (editor). Boston: Kluwer-Nijhoff.

Klein, F. (1918). Über die Differentialgesetze für die Erhaltung von Impuls und energie in der Einsteinschen gravitationstheorie. *Nachr. Akad. Wiss. Göttingen, Math-Phys, KI, II*, 171-189.

Lie, S. (1891). *Vorlesungen über Differentialgleichungen, mit bekannten infinitesimalen Transformationen*. G. Scheffers (editor). Leipzig: Teubner, reprinted 1967, Chelsea Publishing Co., New York.

Liviatan, N. and Samuleson, P. A. (1969). Notes on turnpikes: Stable and unstable. *Journal of Economic Theory*, 1454-1475.

Logan, J. D. (1977). Invariant variational principles. *Mathematics in Science and Engineering, vol. 138*. New York: Academic Press.

Lucas, R. E. (1967). Optimal investment policy and the flexible accelerator. *International Economic Review* (February 8).

Maeda, S. (1980). Canonical structure and symmetries for discrete systems. *Math. Japon, 25*, 405-420.

Maeda, S. (1982). Lagrangian formulation of discrete systems and concept of difference space. *Math. Japon, 27*, 336-345.

Maeda, S. (1987). Completely integrable symplectic mapping. *Proc. Japan Academy, 63A*, 198-200.

Maeda, S. (1988). Quadratic conservatives of linear symplectic Systems. *Proc. Japan Academy, 64A*, 45-48.

Noether, E. (1918). Invariante variantionsprobleme. *Nachr. Akad. Wiss. Göttingen, Math-Phys, KI, II*, 235-257. Translated by M. A. Tavel (1971). Invariant variation problems. Transport Theory and Statistical Physics, 1, 186-207.

Ramsey, F. (1928). A mathematical theory of saving. *Economic Journal, 38*, 543-559.

Samuleson, P. A. (1967). A turnpike refutation of the golden rule in a welfare-maximizing many-year plan. *Essays on the Theory of Optimal Economic Growth.* K. Shell (editor).M.I.T. Press.

Samuleson, P. A. (1970a). Law of conservation of the capital-output ratio. Proceedings of the National Academy of Sciences, Applied Mathematical Science, 67, 1477-1479.

Samuelson, P. A. (1970b). Two conservation laws in theoretical economics. Cambridge, Massachusetts: M.I.T., Department of Economics mimeo (July).

Samuelson, P. A. (1972). The general saddlepoint property of optimal-control motions. *Journal of Economic Theory, 5*, 102-120.

Samuelson, P. A. (1976). Speeding up of time with age in recognition of life as fleeting. *Evolution, welfare and time in economics: Essays in honor of Nichols Georgescu-Roegen.* A. M. Tang et al. (editors). Lexington, Massachusetts: Lexington-Heath Books.

Samuelson, P. A. (1982). Variations on capital/output conservation laws. Cambridge, Massachusetts: M.I.T., Department of Economics mimeo (January).

Samuelson, P. A. and Solow, R. M. (1956). A complete capital model involving heterogeneous capital goods. *Quarterly Journal of Economics, 70.*

Sato, R. (1981). *Theory of technical change and economic invariance: Application of Lie groups.* New York: Academic Press.

Sato, R. (1982). Invariant principle and capital/output conservation laws. Providence, Rhode Island: Brown University working paper No. 82-8.

Sato, R. (1985). The invariance principle and income-wealth conservation laws: Application of Lie groups and related transformations. *Journal of Econometrics, 3*, 365-389.

Sato, R. and Maeda, S. (1987). Local conservation laws of the discrete optimal growth model. Mimeo.

Sato, R., Nono, T. and Mimura, F. (1983). Hidden symmetrics: Lie groups and economic conservation laws. *Essays in honor of Martin Beckmann.*

Sato, R. and Ramachandran, R. (1987). Factor price variation and the Hicksian hypothesis: A micro economic approach. Oxford Economic Papers, forthcoming.

Weitzman, M. L. (1976). On the welfare significance of national product in a dynamic economy. *Quarterly Journal of Economics, 90,* 156-162.

CHAPTER 7

Choice as Geometry

Thomas Russell

1. INTRODUCTION

The purpose of this paper is to provide a differential geometric framework for the analysis of individual choice under uncertainty. Since this approach is far from standard, we begin by discussing informally how choice can be geometrised, and why we believe the approach has value.

The archetype choice problem with which we shall be concerned has the following structure. An individual is presented with a set of risky prospects which can be viewed as a family of univariate random variables with density functions $f(x, \theta)$ where θ is an n dimensional label or parameter which varies smoothly across the family.

Many choice problems of interest in economics have this structure. We list three.

1. **Portfolio Choice.** An individual must choose shares of wealth to invest in each of k assets, $X_1, X_2, \ldots X_k$. Here the parameter $\theta = (\theta^1, \theta^2, \ldots 1 - \sum_i^{k-1} \theta^i)$ is the share of wealth, and the density family is the density function $f(x, \theta)$ of the asset $\sum_i \theta^i X_i$.

2. **Insurance Policy Design.** The terms of an insurance contract can be altered by changing the levels of deductible and/or co-insurance. This has the effect of generating a family of density functions $f(x, \theta)$ over wealth, where θ parametrises the terms of the contract.

3. Product Safety. By changing the design of a product, the manufacturer can change the risk to health as measured by the dollar costs of medical care. This again leads to a family of density functions $f(x, \theta)$ where now θ parametrises features of the design of a product.

Whatever interpretation we give to the family $f(x, \theta)$ we take it that the purpose of a theory of choice is to predict which $f(x, \theta)$ the individual will choose in the allowable set.

Differential geometry enters the picture in two ways. In the first place, given some mild regularity conditions, a family of density functions $f(x, \theta)$ can be given the structure of a smooth n dimensional manifold, i.e., an object which locally looks like n dimensional Euclidean space.

This idea was introduced into statistics by Rao [1945]. It was extensively developed by Centsov [1972] and has been recently used by Efron [1975], Dawid [1975], Atkinson and Mitchell [1981], and Barnsdorf-Nielsen [1978] among others. Perhaps the most accessible treatment is the monograph of Amari [1985] and it is his treatment which we follow.

The fact that the choice set is a smooth manifold permits us to give geometric content to a number of operations, most importantly the operation of taking the derivatives of functions defined on the manifold. Since choice is frequently viewed as maximising some valuation function (more strictly valuation functional) of density functions (for example, expected utility)

$$\int u(x) f(x, \theta) \, dx$$

the derivative of such functions (for example, marginal expected utility)

$$\frac{\partial \int u(x) f(x, \theta) \, dx}{\partial \theta_i}$$

is a well defined geometric object.

This permits some notational economies in describing first order conditions, but it hardly in itself warrants incurring the switching costs necessary to cope with a new mathematical technique. Yet if the only information that we have about the choice set is that it is a smooth manifold, nothing much further can be said. A smooth manifold has a local Euclidean structure, but that is all.

If we wish to say more about the choice set, we must give it more geometric structure, for example curvature, but where is this additional structure to come from? Here differential geometry enters the story in a much more subtle way. We will argue that the smooth manifold, the choice set, has a shape which is determined by the laws of choice themselves. To make this more concrete, consider how we might generate a choice path on the manifold.

Starting from some initial "point" $f(x, \theta_0)$, ask the individual to indicate the direction in which he would choose to move if he was free to move a small amount $d\theta$. Call this direction 'v'. Repeat this experiment and ascertain a new preferred direction, say 'w'. Integrating these directions then gives a path on the manifold, the "choice path," with the property that the derivative of the path is in the direction of the individual's local chosen direction of movement at each point.

Suppose that the direction of movement changes as we move along the path. How could we account for this in mathematical terms? There are two possible approaches.

1. **Preference Fields**. It could be argued that there is a "preference field" on the manifold, where the role of the preference field is precisely to tell the preferred (and therefore chosen) local direction of movement. The changes in direction on the choice path are then explained by the changes in preference field from point to point.

2. **Curved Manifolds**. Alternatively it may be argued that there is no field. The individual always moves in what is to him the same direction, but the manifold is curved. The changes in direction reflect this curvature.

By analogy, consider an aircraft flying in a straight line at a constant height above the earth over terrain which has hills and valleys. The shadow of this airplane will be seen by observers in the airplane to change direction. This change in direction is due entirely to the shape of the earth below the plane. An observer who is actually inside the plane's shadow will feel no change of direction.

A full blooded geometric approach to choice would adopt this latter viewpoint. The changes in direction of choice which are observed would be due to "hills and valleys" in the choice manifold, an idea made precise by the geometric concept of curvature. Natural scientists will recognise here Einstein's "Principle of Equivalence," that a field is equivalent to a geometric shape. By combining Amari et al's manifold structure of families of density functions with the Principle of Equivalence, preference becomes geometry. In this paper we go some distance towards providing such a theory, but stop short of a purely geometric explanation which we hope to provide at some later date.

Now if the geometric ideas offered here added nothing new to our understanding of human action, they might be judged as pleasing aesthetically but inconsequential substantively. The power of the geometric approach, however, is that it is capable of explaining *any* choice path on the manifold.

Not too long ago economists were almost unanimous in their belief that choice under uncertainty conformed to the Bernoulli hypothesis of maximising expected utility. In this case the only task for the theorist was to work out the implications of the theory, and for that existing tools were adequate.

Today, however, many social scientists, particularly those who have looked at the evidence, doubt that behavior under uncertainty satisfies the rather strict conditions of the independence axiom of expected utility maximisation. Consequently foundational questions of the theory are again in vogue.

The last few years have seen an explosion of new theorising in this area. The geometric approach unifies this new theorising in terms of a small number of geometric categories. More than this, it shows that the new theories, despite their number, are a very sparse subset of the set of all possible theories which would explain the choice path. Since the evidence on the shape of the choice path is itself based on a very small number of experiments, it is helpful to have a theory rich enough to deal with all the evidence which may come in.

Finally, the geometric approach has something to say to the working econometrician. Whatever theories of choice under uncertainty eventually emerge, demand functions for risky prospects will have to be estimated if the theory is to be of any practical use. Data limitations and problems of estimation will dictate that these demand functions have a simple form.

As has been noted in the work of Sato [1981] and Sato and Ramachandran [1989] for aggregate systems, and Russell [1988c] for non stochastic systems of demand functions, simplicity is frequently associated with symmetry, and symmetry in turn implies the existence of Lie Group structures on the manifold. The geometric approach makes clear that if we are to have simple demand functions, robust under changes in assumptions about preferences, the choice manifold must have a very special shape. Most choice manifolds will not have this shape, but if the choice set is generated by a diffusion process, symmetry will lead to enormous simplification of the theory. This accounts for the famous results of Black and Scholes [1973] on option pricing in continuous time.

2. DIFFERENTIAL GEOMETRY: THE ELEMENTS

There is no royal road to differential geometry, and the remarks in this section cannot substitute for the in-depth treatment of a textbook such as say Spivak [1977]. Nevertheless, the basic ideas of the subject are quite intuitive, and in an effort to aid the translation of the concepts of the mathematician into the ideas of the choice theorist, we provide here a vocabulary of the most relevant geometric terms.

The Choice Set as a Smooth Manifold

Our first task is to show how a choice set consisting of a family of density functions can be treated as a smooth manifold, i.e. a set which locally has the structure of Euclidean space. More formally we have

A set of points N is a *manifold* if each point of N has an open neighborhood which has a continuous 1-1 map onto an open set of Euclidean n space R^n.

This map associates with each point p of N an n-tuple $(x^1(p), x^2(p), \ldots x^n(p))$. These are called the coordinates of the point p, and the manifold is said to be n dimensional.

When two open sets of N, say U and V intersect, there may be two distinct coordinate systems, say $x^1, x^2, \ldots x^n$ for U associated with a map f and $y^1, y^2, \ldots y^n$ for V associated with a map g. The map $g \cdot f^{-1}$, $R^n \to R^n$ provides a mapping of the coordinates x to the coordinates y. N is said to be a *smooth* manifold when $g \cdot f^{-1}$ is smooth (infinitely differentiable) and has a smooth inverse.

Suppose that the choice set of an individual is given by

$$F = f(x, \theta)$$

where $f(x, \theta)$ is a family of univariate probability density function parametrised by θ. Assume that the domain of θ, Θ, is an open subset of R^n. Then under a reasonably broad set of regularity conditions, Amari shows that F can be treated as a smooth manifold with θ as its coordinates.

Here are two examples.

Example One: Normal Densities.

Suppose that by his actions an individual can choose the mean μ and the standard deviation σ of a density function from a family of normal density functions so that

$$F' = f(x, \theta) = \frac{1}{\sqrt{2\pi\sigma}} \exp - \frac{(x - \mu)^2}{2\sigma^2}$$

where x measures the dollar value of the outcome of some risky project.

In this case we may set $\theta^1 = \mu$ and $\theta^2 = \sigma$ and the parameter space is the subset of R^2

$$\Theta = (\mu, \sigma), \quad -\infty < \mu < \infty, \quad 0 < \sigma$$

Example Two: States of Nature

To be able to treat the choice set as a manifold we require smoothness in the parameter, θ, not smoothness in the value of the outcome x. Thus families of discrete random variables, such as those which arise in the "state preference" approach to choice under uncertainty, can also be viewed as smooth manifolds provided we think of probability density functions as functions in the generalised or Dirac sense.

As we have shown elsewhere, the Dirac delta function and its derivatives have an important role to play in modern financial economics, Russell [1988d], so we recall their main features here.

The Dirac delta function $\delta(x)$ is a "function" with the property

$$\int \delta(x) G(x) \, dx = G(0)$$

for suitably smooth functions G. Thus the delta function is not a function in the sense of an object with point values, but is rather an operator on other regular functions. It can be thought of as the limit of a family of true functions. Lighthill [1978], for example, shows how the delta function arises as the limit of a family of Gaussian functions with zero mean as the variance becomes smaller. That is

$$\delta(x) = \lim_{n \to \infty} \exp^{-n x^2} (n/\pi)^2$$

where $n = 1 / 2 \, \sigma^2$.

Figure 1 shows members of this family for various values of n. The important properties of the delta function which we use are

1: Shift Property

$$\int \delta(x-a) \, G(x) \, dx = G(a).$$

Here $\delta(x-a)$ can be thought of as the probability one spike at the point a. In this way discrete probabilty density functons become a row

Figure 1. The Dirac Delta Function as a Limit.

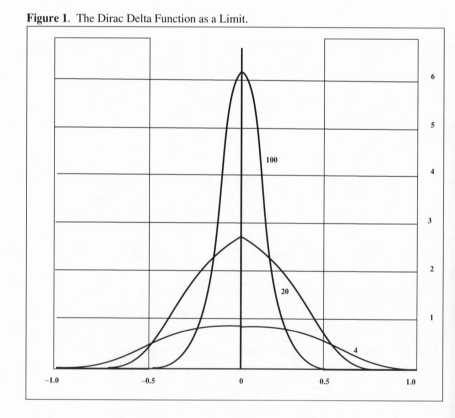

of delta functions each $\delta(x-a)$ being weighted by the probability of the event a, $p(a)$.

2: Derivative Property

The nth derivative of the delta function, $\delta^{(n)}(x)$ is given by

$$\int \delta^{(n)}(x)\, G(x) = (-1)^n\, G^{(n)}(0)$$

For example, the first derivative of the delta function can be thought of as the limit of the derivatives of the family of Gaussians in Figure 1, Figure 2.

Figure 2. The First derivative of the Dirac Delta Function as a Limit.

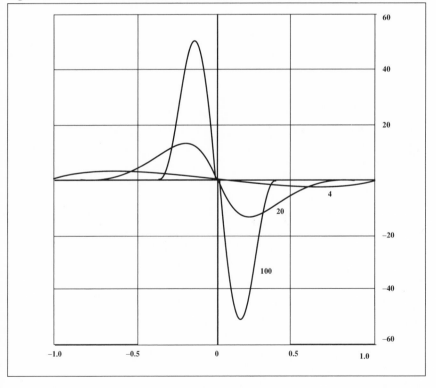

Derivatives of delta functions play an important role in analysing the manifold of a continuous time diffusion à la Samuelson [1965] [see Black and Scholes (1973)].

Let there be three states of nature [see e.g. Debreu (1959)] s_1, s_2, s_3, with probabilities p_1, p_2, p_3 where $p_3 = 1-(p_1 + p_2)$. Let the payoff to an individual if state s_i occurs be a_i. Then the density functions of the outcomes is given by

$$f(x, p) = p_1 \delta (x-a_1) + p_2 \delta (x-a_2) + p_3 \delta (x-a_3)$$

Suppose that an individual through his actions can influence p_i. Then the choice set becomes

$$F = f(x, \theta) = \theta^1 \delta (x-a_1) + \theta^2 \delta (x-a_2) + \theta^3 \delta (x-a_3)$$

where we let $\theta = p_i$ and the parameter set is

$$\Theta = \theta^i \mid \theta^i > 0, \ \sum \theta^i = 1$$

In both examples we may treat the choice set as a smooth manifold by using the obvious mapping,

$$\phi : F \rightarrow R^n, \ \phi [f(x, \theta)] = \theta$$

to give coordinates to each point of F. These coordinates are not unique. Any other set of coordinates, say ρ, can also be used to label the points of F provided $\rho(\theta)$ and $\theta(\rho)$ are one to one and smooth. Sometimes, as we shall see, special coordinates will play special roles in simplifying the theory.

Tangent Vectors, One-Forms, and Their Spaces

Suppose, then, that N is a smooth manifold. A smooth manifold has a well defined local geometric structure. To make this structure precise without using coordinates, the local structure (the tangent approximation) is described using the derivatives of functions from R^1 to the manifold. Since the derivative of such a function gives a local direction of movement, tangents can be identified with directional derivative operators.

Figure 3 shows the steps to defining a tangent vector at θ_0. Step one is to define a mapping from R^1 to N, and take its derivative, $\bar{v}$. Step two is to define the tangent vector as the equivalence class of the derivatives of *all* mappings from R^1 to N which have $\bar{v}$ as their tangent at θ.

More formally, define a *curve* as a mapping from an open set of R^1 into N. If we use λ to parametrise this open set of R^1, we can in this way associate points θ in N with λ. Now suppose we have a curve through the point θ_0 of N given by $\theta = \theta(\lambda)$. If we have a differentiable function on N, say f, the value of this function on N sets up a mapping on the curve from R^1 to R^1 which we write as

Figure 3. Tangent Vectors as Directional Derivative Operators.

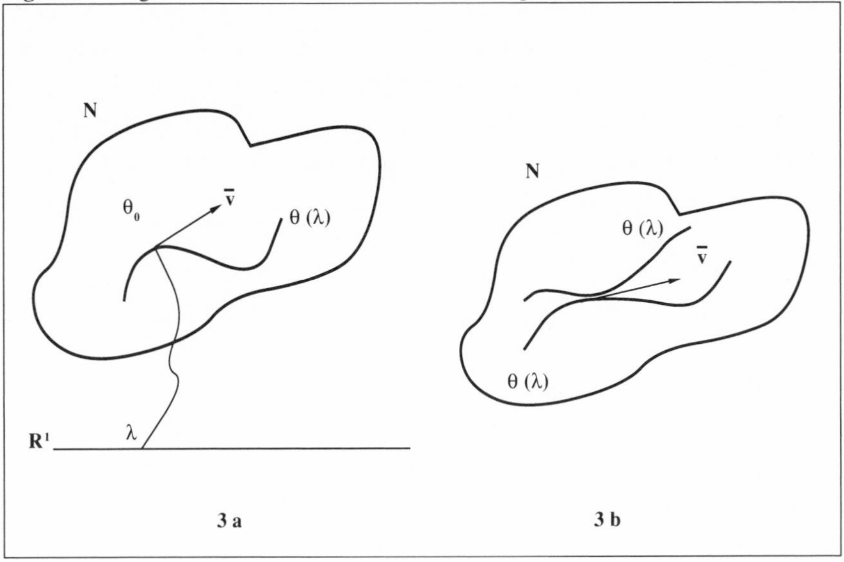

$$g(\lambda) = f(\theta^1 (\lambda), \ \theta^2 (\lambda), \ \dots \ \theta^n (\lambda)).$$

Differentiating we have

$$\frac{dg}{d\lambda} = \sum \frac{d\theta^i}{d\lambda} \frac{\partial f}{\partial \theta^i}$$

This will hold for any function f. So we call

$$\frac{d}{d\lambda} = \sum \frac{d\theta^i}{d\lambda} \frac{\partial}{\partial \theta^i}$$

a tangent vector to the manifold at θ_0. Many curves will have $d / d\lambda$ as their tangent. The tangent vector is thus an equivalence class of derivatives along curves. We show two members of this family in Figure 3b.

The set of all tangent vectors to a manifold at the point θ_0 is called the tangent vector space. It is written T_{θ_0}. We will write vectors in T_{θ_0} with a bar, i.e. $\bar{v}$. The space of all vectors at all points is also a smooth manifold. It is called the tangent vector fiber bundle. The fibers are the tangent spaces. The selection of one point in the fiber at each point in the manifold is called a section of the fiber bundle. A vector field is a rule which allocates a vector to every point in a manifold. Thus a vector field is a section of the tangent vector fiber bundle.

Associated with the tangent vectors in T_{θ_0} are their duals, the linear operators on vectors called one-forms or covectors. More formally, a one-form is a linear real valued function of vectors. Perhaps the most commonly used one-form is the gradient of a function. It can be shown that the space of one-forms at θ_0 is also a vector space. It is written as $T^*_{\theta_0}$ Just as with vectors, the space of all covectors at all points is a manifold, the cotangent fiber bundle. A covector field is a section of the covector fiber bundle.

If we have a mental picture of a vector as an arrow, then the equivalent mental picture of a one-form must be a set of equally spaced parallel lines, Figure 4. The number of lines crossed by a vector is the value of the vector on the one-form. To an economist, a one-form is the local and therefore "straightened out" approximation to an indifference curve. Speaking loosely, if a vector field *describes* behavior on a manifold, a one-form field *prescribes* it.

Figure 4. Tangent Vectors and One-Forms.

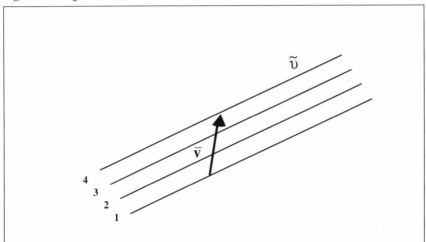

Representations of the Tangent and Cotangent Space

When a set of coordinates is given, the coordinate functions themselves are curves on N. In this case, the derivatives of these coordinate functions $\partial / \partial \theta^i = \partial_i$ provide a basis for T_θ and any vector in T_θ can be written as $\sum A^i \partial_i$ or more succinctly as $A^i \partial_i$ where we use the Einstein summation convention that indeces raised and lowered are summed.

There is, however, a more natural representation of the tangent space from the point of view of decision making under uncertainty. Let

$$l(x, \theta) = \log f(x, \theta)$$

and consider the n partial derivatives

$$\frac{\partial \log f}{\partial \theta^i} = \partial_i l \qquad i = 1, 2, \ldots, n$$

Provided these are linearly independent, we can construct an n dimensional vector space called by Amari T_θ^l (the one-representation), where

$$T_\theta^1 = (A(x) \mid A(x) = A^i \, \partial_i \, l(x, \theta))$$

T_θ^1 is thus the linear space of random variables spanned by $\partial_i l(x, \theta)$. Now make the obvious association

$$\partial_i \in T_\theta \Leftrightarrow \partial_i l(x, \theta) \in T_\theta^1.$$

This associates a directional derivative operator $A = A^i \partial_i \in T_\theta$ with a random variable $A^i \partial_i l(x, \theta) \in T_\theta^1$. It is easy to check that all the random variables in T_θ^1 have zero expectation under $f(x, \theta)$. If the manifold has a boundary, then the tangent space may fail to have full dimension on the boundary. The dimension of the tangent space on boundaries is well discussed in Pfanzagl [1982], Chapter 1, who calls such degenerate tangent spaces tangent cones.

The cotangent space associated with T_θ^l, T_θ^{1*} is the space of linear operators on the random variables in T_θ^1. Suppose, for example, that we have some valuation functional $V(f(x, \theta)$ which associates a real number with the density functions in N. Then if V has a gradient, that gradient is a member of T_θ^{1*}, see Pfanzagl [1982], Chapter 4, i.e.

$$\text{Grad } V = \tilde{u}$$

where tilde indicates that the object is a one-form.

If an individual acts to maximise some smooth functional on N, the first order conditions at θ_0 have the simple coordinate free representation

$$\tilde{u}(\bar{v})(\theta_0) = 0 \qquad \text{for all} \qquad \bar{v} \in T_{\theta_0}^1$$

Here we see the advantage of using the $(1-)$ representation of the tangent space. Often in studying choice problems we are interested in studying separately

(a) the effect of a change in the family of random variables which give the choice set, and

(b) the effect of a change in preferences. By using the $(1-)$ representation the choice set is described locally by a vector, and preferences are described locally by a one-form.

Differentiating Vectors 1: External Differentiation and the Hodge Star Operator

So far the discussion in this paper has focussed entirely on the local properties of the manifold. To describe the behavior of tangent vector and cotangent vector fields, we need to know how to take the derivative of these objects on a manifold. This, in general, requires additional structure.

There is one derivative operation, however, which can be taken with only the geometric equipment we already have. Suppose that we call a function a 0-form. Then the operation of taking the gradient of the function takes a 0-form and produces a one-form. This operation, which raises the order of the form by one, can be undertaken on forms of any order and is called exterior differentiation. It is written d, so that df is a one-form, the gradient of the function f. On a one-form, say $\tilde{u}$, d produces a two-form. A two-form is an anti-symmetric linear real valued function of two vectors. Thus if W is a two-form

$$\Omega\,(\bar{v}, \bar{w}) = -\Omega\,(\bar{w}, \bar{v})$$

Intuitively a two-form measures the local rotation of a one-form field, and is therefore a local test of intransitivity. The mechanics of taking the exterior derivative can be found in any textbook.

Associated with exterior differentiation is the so called Hodge Star operator $\star$. The $\star$ operator sets up a map between p forms on an n dimensional manifolds and n-p forms on the same manifold. Thus if $\tilde{u}$ is a one form on an n dimensional space, $\star\tilde{u}$ is an $n{-}1$ form. $d\,\star\tilde{u}$ is then an n form, and thus $\star d\,\star\tilde{u}$ is a zero form. The details of how to calculate $\star$ can be found in the textbooks. (Very) roughly, $\star d\star\tilde{u}$ calculates how a one-form field is splaying out. To calculate $\star$ we need to have a metric on N. As Amari shows, the Fisher Information Matrix g will provide a metric where

$$g_{ij}(\theta) = \,<\partial_i, \partial_j> \,= E[\partial_i\, l(x, \theta)\, \partial_j\, l(x, \theta)]$$

Differentiating Vectors 2: Covariant Differentiation and the Connection

There is another way to form the derivative of vectors (and therefore one- forms) on a manifold. This requires a connection, a different, and weaker, geometric structure than a metric. The derivative operation made possible by a connection is called the covariant derivative.

The purpose of a connection, as its name implies, is to link or connect tangent spaces. To do this in coordinates it is necessary to specify how the ith coordinate of the jth tangent vector is changing in the kth direction. On an n dimensional manifold this will require the specification of n^3 functions of θ.

These functions, which we write as $\Gamma_{ijk}(\theta)$, are called the Christof-fel symbols or connection coefficients. On a manifold without further structure, the Christoffel symbols can be chosen arbitrarily. However on a manifold of probability density functions, some restrictions must be applied to make sure that each tangent space is a space of random variables with zero expectation.

The class of allowable connections was first given by Centsov for discrete density functions. They are called by Amari the α connections. Two members of this class have special significance for the theory of choice under uncertainty. We have

1: The Mixture (or Dawid) Connection

$$\Gamma_{ijk} = E[(\partial_i \partial_j l(x, \theta) + \partial_i l(x, \theta)\partial_j l(x, \theta)\partial_k l(x, \theta)]$$

With this connection spaces of mixtures of probability density functions become flat spaces and can therefore be given a coordinate system in which the Christoffel symbols disappear.

2: The Metric (or Riemannian) Connection

$$\Gamma_{ijk} = \frac{1}{2} [\partial_i g_{jk} + \partial_j g_{ik} + \partial_k g_{ij}]$$

where g_{ik} is the ijth entry in the Fisher Information Matrix defined previously. This connection is the only symmetric connection compatible with the metric introduced previously.

The covariant derivative of a vector $\bar{v}$ in the direction $\bar{w}$ is written $\nabla_{\bar{w}}\bar{v}$. It is given by

$$\nabla_{\bar{w}}\bar{v} = (\frac{dV^j}{d\gamma} + \Gamma_{ki}^j V^k U^i) \bar{\varepsilon}_j$$

where $U = d / d \lambda$ and $\bar{\varepsilon}_j$ is a basis vector. Without formally dealing with the issue of curvature of a manifold, we may think of the covariant derivative as the directional derivative on a curved manifold which corrects the partial derivatives for the curvature of the manifold. Thus if $\nabla_{\bar{v}} \bar{v} = \lambda \bar{v}$, i.e., the covariant derivative of a vector along the path of the vector points in the direction of the vector, then the path of the vector is travelling as straight as it can. This property defines a geodesic on the manifold. Note that which paths are geodesics depends on the choice of connection.

3. CHOICE AS GEOMETRY

In this section we present theories of choice in geometric terms. The basic idea is to treat local choice as though it is governed by a preference field. The geometric object which corresponds to this preference field is either a one-form field (i.e., a cross-section of the co-tangent bundle) or a two-form field.

In principle, these fields could be mapped out by a carefully designed set of experiments. The pioneering work of Allais [1952] and the recent work of Weber and Camerer [1987] and Camerer [1988] represent important advances in this direction. If such a map can be obtained, then the problem of the theorist is to describe it in succinct terms, as Maxwell did for Faraday's map of the electromagnetic field. We can reasonably require that this description not depend on the coordinates used to describe the manifold, i.e., that it be in covariant form.

This approach does not require any axioms of choice. Frequently models of choice are written in axiomatic terms, for example the Von Neumann, Morgenstern axioms. The axioms, however, are seldom of any practical use, and are usually used to deduce the existence of a preference functional, for example, $V = \int U(x) f(x, \theta) \, dx$. The individual's choice is then assumed to maximise this functional on the choice set. The first order conditions for this maximisation require knowledge of the gradient of V. But the gradient of V is a one-form. Thus the geometric theory begins at the point at which the axiomatic theory confronts the data.

Suppose, then, that we have a pair $(N, \tilde{u})$, where N is a smooth n dimensional manifold (with coordinates θ) and $\tilde{u}$ is a one-form field on N. We view this field as providing the impetus for choice under the following hypothesis.

Hypothesis of Choice 1. If at any point θ in N, there exists a vector $\bar{v}$ such that $\tilde{u}(\theta)\ \bar{v}(\theta) > 0$, then the individual will choose to trade in the direction v. An equilibrium choice is a point θ where $\tilde{u}(\theta)\ \bar{v}(\theta) \leq 0$ for all $\bar{v}$ at θ. [1]

Accepting this hypothesis for the moment, we face the task of describing $\tilde{u}$. $\tilde{u}$ is a first order object on N. To describe how the field behaves locally we need some way of calculating the derivative of $\tilde{u}$, a second order object. Unfortunately there is no canonical way to take the derivatives of forms. Put differently, the fact that N is a smooth manifold does not in itself provide enough structure to identify a constant one-form on N.

To do this we need extra structure and this can be given in at least two different ways, via the external derivative operator d or via the covariant derivative operator ∇.

Exterior Derivation. It is well known to physicists that in three-dimensional Euclidean space a vector (or one-form) field is completely characterised by two operations on the field, "Div" and "Curl." One may think of the electrostatic field E for which

$$\text{Div } E = 4\,\pi\,\rho$$
$$\text{Curl } E = 0$$

or the magnetostatic field B for which

$$\text{Div } B = 0$$
$$\text{Curl } B = 4\,\pi\,J$$

Although Div and Curl are defined for vectors on three-dimensional Euclidean space, they both have generalisations for one-form fields on smooth manifolds of arbitrary finite dimensions, the situation we face here. Curl $\tilde{u}$ generalises to $d\tilde{u}$ the exterior derivative of $\tilde{u}$, Div $\tilde{u}$ to $\star d$ ($\star\tilde{u}$) where $\star$ is the Hodge Star operator.

To compute $d\tilde{u}$ requires only that the manifold be smooth. As we shall see in a moment, $d\tilde{u}$ provides information about the transitivity of preferences. To compute $\star d$ ($\star\tilde{u}$), however, we need a metric on N. This is provided by the Fisher information matrix

$$g_{ij}(\theta) = <\partial_i, \partial_j> = E[\partial_i \, l(x, \theta) \, \partial_j \, l(x, \theta)].$$

Thus one way to describe the preference field is write down the equivalent of Maxwell's Laws for the field, i.e.

$$d\tilde{u} = \Omega \, (\theta)$$
$$\star d \, (\star\tilde{u}) = \phi \, (\theta)$$

for some functions ϕ and two-form Ω.

Covariant Derivation. The second way to proceed describes the field by its covariant derivative, $\nabla \, \tilde{u}$. To do this we need a connection on N. At this point it becomes easier to see what is happening if we specialise the manifold. We shall consider M, a manifold of mixtures of probability density functions. Let $f_1, f_2, \dots f_n$ be n arbitrary non-equal probability density functions, and let

$$M = \theta^1 f_1 + \theta^2 f_2 + \dots \theta^n f_n$$

where $\theta^i \geq 0 \sum \theta^i = 1$. M can be thought of as the unit simplex in R^n without its linear structure.

A special case of this manifold has been studied by Marschak [1950], Machina [1982], Samuelson [1983] and Weber and Camerer [1987], among others. Let

$$T(\theta) = \theta^1 \, \partial(x-a_1) + \theta^2 \, \partial(x-a_2) + \theta^3 \, \partial(x-a_3)$$

with $a_1 > a_2 > a_3$ and $\theta^1 > \theta^2 > \theta^3$. Choosing as coordinates for R^2, θ^1 and θ^3 we can think of T as a triangular subset of the first quadrant of R^2 with its linear structure removed.

Suppose that we choose as a connection on M the mixture connection with Christoffel symbols

$$\Gamma_{ijk} = E\,[(\partial_i\,\partial_j\,l(x,\,\theta) + \partial_i\,l(x,\,\theta)\,\partial_j\,l(x,\,\theta))\,\partial_k\,l(x,\,\theta)]$$

Then we know from Amari that M and T are flat spaces. Moreover the coordinates θ have the special property that the Christoffel symbols are zero globally in these coordinates. This means that to take a covariant derivative Δ we need simply compute the partial derivative ∂. Thus we can describe the various fields on M by giving the partial derivatives of their coefficients and be certain that we are describing the field in a coordinate free way. The description, however, does depend on the choice of connection.

Even without a connection we can still take the exterior derivative du. There are two possibilities, $d\tilde{u} = 0$, the field is closed, $d\tilde{u} \neq 0$, the field is not closed; see Figures 5A (closed) and 5B (non-closed).

If $\tilde{u}$ is not closed at θ, then integrating the field around a closed path will not produce the value zero. (Such a field is sometimes called rotational.) We do not view this as fatal to a positive theory of choice, but instead will place this type of field to the side for the moment and concentrate on describing closed one-form fields.

CLOSED ONE-FORM FIELDS

If the field u is closed, we can apply Poincare's Lemma.

Poincare Lemma. *A closed one-form field defined on a region which can be deformed to a single point is exact.*

That is, if $d\tilde{u} = 0$, $\tilde{u} = dV$ where V is a zero form, in this case a functional.

Figure 5. Closed and Non-Closed One-Form Fields.

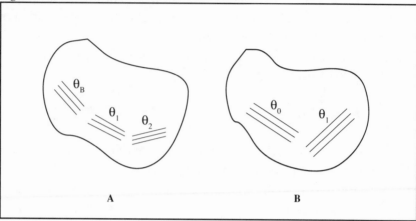

Since the manifold M has no holes, the topological prerequisites of the Lemma are satisfied. Thus, by integration, we can obtain a functional V with the property that $\tilde{u} = dV$, i.e., $\tilde{u}$ is a gradient field. Rather than describing the field by integration techniques, however, we can instead describe it by describing its local variations, i.e., by differential techniques.

The simplest field we can consider is the constant field on M, i.e.,

$$\nabla_{\bar{v}} \tilde{u} = 0 \qquad \text{for all} \quad \bar{v}$$

This field is generated by taking its value at any point in M and moving the field around so that it remains parallel and equally spaced at all points, Figure 6.

But this is clearly the field of an expected utility maximiser. If we express $\tilde{u}$ in coordinates $u_i(\theta)$, then a constant field has the property that $\partial u_i / \partial \theta^j = 0$ for all i, j. If we perform the line integral

$$\int_{\theta^2}^{\theta^1} \tilde{u}$$

Figure 6. The Field of an Expected Maximiser.

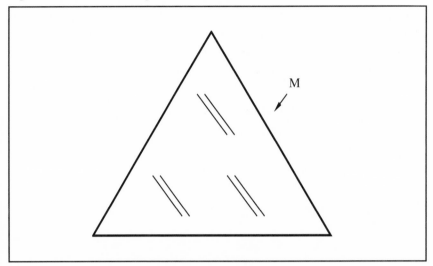

then by Stokes Theorem of differential calculus, this integral is equal to the difference of the values on its endpoints, i.e.

$$V(f(x, \theta^2)) - V(f(x, \theta^1))$$

It is also clear by differentiation that the functional

$$V(f(x, \theta)) = EU = \sum \theta^i \int u(x) f_i(x) \, dx$$

satisfies the property that the one-form dV is constant on M. We can thus view expected utility maximisation as the theory of choice which produces a covariant constant one-form preference field on M.

This point of view has a number of consequences. In the first place, we see that there is an intimate link between the curvature of mixture space as determined by the connection on it and the description of the preference field. By choosing the mixture connection, we make expected utility fields constant. At the same time we make mixture space flat. By choosing another connection we could make some other one-form field covariant constant. But then we make mixture space curved. We will return to this idea when we introduce other theories of choice. For the moment we take the more old fashioned view, and retain the mixture connection on *M* describing the field in terms of their covariant derivatives on *M*.

Since *u* is a constant field, we must have

$$\frac{\partial^2 V}{\partial \theta^{i\,2}} = 0$$

This in turn means that

$$\sum \frac{\partial^2 V}{\partial \theta^{i\,2}} = 0$$

or in covariant terms

$$\sum \nabla_{\partial \theta^i} u^i = 0$$

where u^i is the *i*th coordinate of *u* in the dual basis. In a Euclidean space with rectangular coordinates, the operator

$$\Delta = \sum \frac{\partial^2}{(\partial \theta_i)}$$

is known as the Laplacian, and the Laplacian can be used to describe and recover the field through Poisson's equation $\Delta = \phi\,(\theta)$. The generalisation to arbitrary coordinates is known as the operator Div Grad and this can be generalised to arbitrary manifolds if we have a metric. We can also calculate the Laplacian of $\tilde{u}$ by taking the covariant divergence. However, since the mixture connection is not a metric connection, the covariant Laplacian will differ from the Laplacian defined in terms of the metric.

If the data is not consistent with a constant field, then we can try other specifications. For example, we may propose

$$\nabla_{\partial \, \theta^i} \, \tilde{u}^i = k \qquad k \text{ a constant for each } i$$

This would be generated by a functional which is quadratic in Expected Utility, e.g.,

$$\int u(x) f(x, \, \theta) dx + [\int \tau(x) f(x, \, \theta) \, dx]^2$$

Machina [1982]. Alternatively, an inspection of the evidence may suggest that the field behaves in special ways along special curves in T. For example it may be that $\nabla_{\partial \, \theta^i} u^i = 0$ for each i along rays in T. In this case indifference curves "splay out" in a regular way, Figure 7.

Figure 7. Weighted Expected Utility.

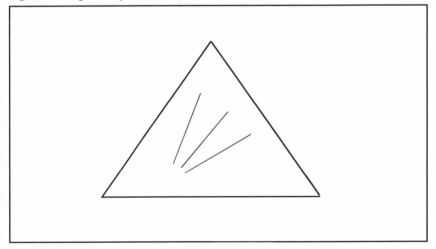

This behavior would be generated by a functional of the form

$$V = \frac{\int u(x) f(x, \theta) \, dx}{\int \tau(x) f(x, \theta) \, dx}$$

sometimes called Weighted Expected Utility, Chew and Macrimmon [1979] Chew [1983].

Any behavior which can be represented by local indifference curves on T can of course be specified by specifying $\nabla \tilde{u}$. Thus the theories of Rank Dependent Utility, Quiggin [1982], Yaari [1987], Disappointment Theory, Loomes and Sugden [1986], Lottery Dependent Utility, Becker and Sarin [1988], and Prospect Theory, Kahnemann and Tversky [1979], can all be specified in terms of $\nabla \tilde{u}$. Note that in all of these theories there is some connection in which the field is constant. With that connection mixture space will be curved. But since it is smooth, it is locally flat. Therefore it is not possible to distinguish among closed one-form theories of choice by observations confined to the tangent space. It seems particularly apt therefore to call closed one-form theories General Expected Utility, since in the natural sciences General Relativity is precisely the curved space form of Special Relativity and the same inability to distinguish the theories locally applies.

Weber and Camerer (1987) and Camerer (1987) have analysed these theories and others by looking at the slopes of the indifference curves in T. From the geometric point of view they are implicitly using the mixture connection on T. By making the connection explicit, we derive covariant expressions which are valid in any coordinates on any choice manifold (with connection) with any dimension and which allow any *a priori* choice behavior consistent with local indifference curves.

Non-Closed One-Form Fields

Non-closed one-form fields have the property that $d\tilde{u} \neq 0$. The exterior derivative of $\tilde{u}$ is then a two-form field Ω. What happens when a one form field is not closed? In this case the local one-forms do not mesh together to form a system of indifference curves. As a consequence of this, if we make a closed vector loop in this space and count the number of indifference curves we cross, the net sum is not zero, See Figure 8. (Another way to say this is that the value of integrals is not path independent.) In Figure 8, if we go from 0 to a to $a + b$ to b to 0 we cross net +3 indifference curves. (Note that the indifference curves have an orientation so that we know what is an increase in utility and what is a decrease.)

The net number of indifference curves crossed by a closed vector loop is generated by the two-form. Just as a one-form takes a single vector and produces a real number in a linear way, so a two-form takes two vectors and produces a real number in a linear way. If in Figure 8 we traverse the closed loop in the opposite direction we will cross the same number of indifference curves, but with the opposite sign. Thus a two-form defined in this way is anti-symmetric, i.e., $V(a, b) = -V(b, a)$.

The geometric object which represents a two-form in its own right is a pattern of cells (like the inside of a box of wine) with a sense of direction marked on the cells. In two dimensions, if we take two vectors and generate a plane by completing the vector parallelogram, the two-form gives the net number of cells which these two vectors pierce. See Figure 9.

Suppose, then, that we have an arbitrary choice manifold, N on which is defined a non-closed one-form field $\tilde{u}$ with $\Omega = d\tilde{u}$. The hypothesis of choice implicit in a two-form theory is as follows.

Figure 8. The General One-Form Field.

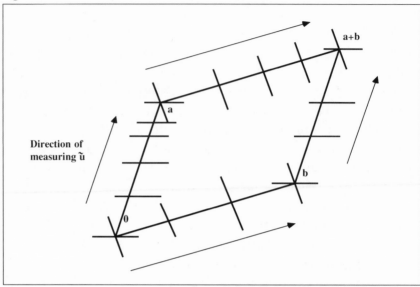

Figure 9. The Two-Form.

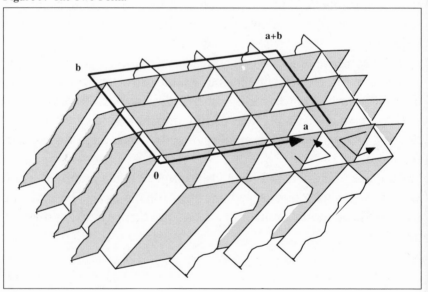

Hypothesis of Choice 2. *If at any point* θ *in* N *there exists two vectors* $\bar{v}$ *and* $\bar{w}$ *with* $\Omega(\bar{v}, \bar{w}) > 0$, *then the individual will choose to trade in the direction* w *rather than the direction* $\bar{v}$. *An equilibrium choice is a point* θ^0 *at which* $\Omega(\bar{v}, \bar{w}) < 0$ *for all* $\bar{v}, \bar{w}$ *in* θ^0 *with equality if* θ^0 *is an interior point of* N.

Something like this hypothesis has recently been proposed by a number of authors, e.g., Bell (1982), Loomes and Sugden (1982) and Fishburn (1984) under the general title "Regret Theory." In Regret Theory if $f(x)$ and $f(y)$ are two density functions, then an asymmetric function $r(x, y)$ is proposed and $f(x)$ is chosen over $f(y)$ if

$$R(f(x), f(y)) = \int\int r(x, y) f(x) f(y) \, dx \, dy > 0$$

If we think of some reference lottery, say 0 with certainty, then we may look at R as a surface integral of Ω over the surface bounded by $0, f(x)$ and $f(y)$, just as we earlier viewed Expected Utility as a line integral. Since the two-form Ω is derived by taking the derivative of a one-form, we may call the one-form u a "vector" potential. If we have a single vector, say $\bar{v}$, and value the two-form on this vector, a one-form results, the contraction of Ω on $\bar{v}$. This one-form can now be used to evaluate other vectors, say $\bar{w}$, by placing $\bar{w}$ in the empty slot in Ω.

If we accept Hypothesis 2, then we again face the task of describing the two-form in simple terms. Since the two-form is derived from a one-form we could as easily describe the one-form. But we have already done this. Either give the values of $d\tilde{u}$ and $\star d$ ($\star\tilde{u}$) or give the values of $\nabla\tilde{u}$.

An obvious generalisation of Expected Utilty would be Expected Regret, where the two-form is constant on the mixture manifold M, but there is no need to constrain the two form to be constant. Indeed there is no reason why only one field should be present on M, and a theory of choice could provide for interaction among fields.

Finally we should note that not all two-forms arise as derivatives of one- forms. We could postulate that the prefernce field is a two-form field for which $d\,\Omega \neq 0$. However, Hypothesis 2 would not apply in this case. Here the two-form would serve as potential for a three-form and choice would be based on comparing three choices.

As we can see, there is something of an embarrassment of riches for the theorist who uses geometric methods to describe preference fields. Whenever choice is based on a smooth local evaluation of univariate density functions, however, **all** of these theories can be subsumed under two testable equations

$$\star d\,(\star \tilde{u}) = \phi$$
$$d\tilde{u} = \Omega$$

These equations describe preferences on any manifold of any dimension. The task of the empiricists will be to fill in the right hand side.

4. SINGLE PERIOD PORTFOLIO THEORY AND ASSET PRICING

An important application of the theory of choice under uncertainty is to the theory of asset pricing. Assets, like everything else, are priced so that aggregate demand is equal to aggregate supply. Aggregate supply may be taken as fixed, so that attention usually focuses on aggregate demand.

The demand for an asset by an individual, say the demand for General Motors stock, can be thought of as a component of the demand by the individual for a portfolio of assets. This in turn can be thought of as the choice of a point θ in $P(\theta)$, the portfolio manifold. In order to specify the density function of a portfolio of n assets X_i, we must know the joint probability density function of returns on the assets. The density functions of these original assets, $f(x, i)$, are then interpreted as the marginals of this joint density. The manifold $P(\theta)$

is generated by investing shares of wealth $a^i W = \theta^i$ in the underlying assets to give the density $f(x, \theta)$ at the point θ. $f(x, \theta)$ depends on the joint density function of the underlying assets. In general, given a mixture connection, P will be a curved manifold when embedded in a space of mixtures of density functions. Indifference curves of expected utility maximisers, however, are flats, so that the typical portfolio maximisation problem is the exact obverse of the typical problem of choice under certainty. Here the budget sets are curved and indifference curves are flat.

One possible way to provide coordinates for mixture space is to use moments. Moment space, i.e. a space coordinatised by the moment sequences $\mu_1, \mu_2, \ldots$, is then a countably infinite dimensional space. Obviously this space cannot be renderred in all its dimensions, but schematically the generic portfolio problem looks like Figure 10.

In Figure 10, $f(x, 1)$ and $f(x, 2)$ are two probability density functions of rates of return whose mixture in this three dimensional version of mixture space is the straight line joining them. The portfolio set P is curved in this space.

One of the major problems faced by empirical investigators is that the dimension of P is very large. Thousands of different assets are traded daily on the organised asset markets in the U.S. Add to that international assets and all the non-organised asset markets, and the dimension of P could be in the tens of thousands. Clearly there is a need to reduce this dimension in some way.

Theories of asset pricing typically reduce the dimension of the problem by requiring that asset demands lie in some low dimension submanifold of P. This manifold is usually generated by taking linear combinations of a small number of assets in P.

The aggregation theorems which result from this are known as n-fund portfolio separation theorems, where n is the number of assets in P needed to generate the submanifold. Separation can be produced by preference restrictions, which we call C-S separation (after Cass and Stiglitz [1970]) or by distributional restrictions, which we call R-

Figure 10. The Portfolio Problem.

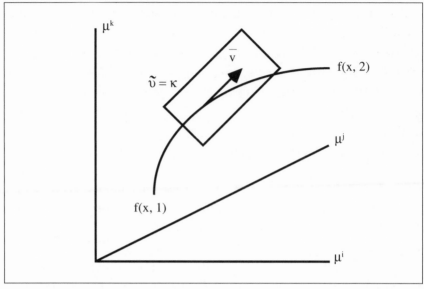

separation (after Ross [1978].)

Geometric tools can be used to provide some insight into these aggregation conditions. We concentrate here on 1 and 2 fund separation. So, let P be a smooth n dimensional manifold parametrised by θ and let $\tilde{u}$ be a closed-one form field on P. $\tilde{u}$ may be the gradient of an expected utilility functional, but need not be.

Case a: R 1-Fund Separation. Suppose that P is one dimensional, i.e., it is generated by two assets. Generalising to n dimensions is straightforward, but with a one dimensional manifold we can draw pictures.

The first order conditions for a maximum at θ^0, $\tilde{u}\,(\bar{v}) = 0$ look like Figure 11a, where we draw the tangent space $T_{\theta^0}^1$. The vector $\bar{v}$ crosses no one-form lines, so the value of the gradient of V in this direction is zero. Now if we rotate $\tilde{u}$ so that we change preferences, (Figure 11b) in general the first order conditions will no longer be satisfied. The only way to prevent this is to have $\bar{v} = 0$. Thus a necessary condition for R 1- Fund Separation is that the dimension of the tangent space of P at θ^0 be zero. This implies that θ^0 is on the boundary of P in some sense. The sense is made clear if we embed P in mixture space using moments as coordinates. $\bar{v} = 0$ implies there is a cusp at θ^0. which we show in 3 moment dimensions in Figure 12. This cusp guarantees that if V is maximised at θ^0 so is V', where V' is any other V.

Case b: C-S 1-Fund Separation. Cass and Stiglitz require that 1 Fund Separation hold when we change the wealth of the investor. Again consider a 1 dimensional manifold P, but now hold wealth constant at say 1 on P so that

$$P = f \mid f = f(x, a^i.1)$$

Now parametrise the one-form field by W so that preferences are given by a form $u(W)$. For example, the Expected Utility functional is given by

$$\int u(x.W) f(x, a) \, dx$$

Here f is the density function of return per dollar invested.

Figure 11. First Order Conditions in Pictures.

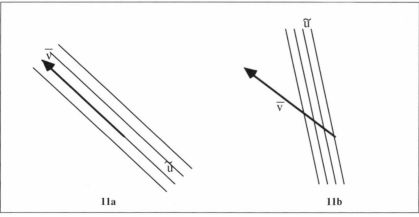

| 11a | 11b |

Now Figures 11a and 11b again apply. If we change W we rotate $\tilde{u}$. In general the vector $\bar{v}$ will not lie on $\tilde{u}$, and the point will not remain a maximum. The only way to prevent this for arbitrary $\bar{v}$ is to stipulate that changes in W "blow-up" $\tilde{u}$, but do not rotate it. That is, $\tilde{u}(W) = h(W)\tilde{u}$ for some function h, i.e., $\tilde{u}$ is homothetic.

For expected utility maximisers it is easy to see that this implies

$$u(x.W) = h(W)\,g(x)$$

the well known condition for constant absolute risk aversion, which solves to give

$$u = aW^b + k.$$

For generalised preferences we must have

$$V(f(x,W)) \;=\; h(W)\,Z(f(x))$$

Case c: R 2-Fund Separation For R 2-Fund Separation we require that optimal choices lie on a path in asset space given by $a\theta^1 + (1-a)$ θ^2 where θ^1 and θ^2 are two points in P. Now let P be two dimensional (generated by three assets.) The typical first order conditions $\tilde{u}(\partial_1 l)$ $= \tilde{u}(\partial_2 l) = 0$ now look like Figure 13, where we have chosen coordinates so that $\partial_2 l$ is in the direction of the optimal path. This is the tangent space at the point θ^0.

Looking at Figure 13 we see that we can generate new forms, and therefore new preferences by rotating along the $\partial_1 l$ axis, or by rotating along the $\partial_2 l$ axis. Unfortunately, we cannot in general assign one instrument, movements along the optimal line, to hit two targets, correct for rotation around each axis. If movements along the optimal path correct for rotations around $\partial_1 l$, then all rotations around $\partial_2 l$ must satisfy the necessary conditions without moving from θ^0. But this implies $\partial_1 l = 0$, i.e., the dimension of the tangent space along the

Figure 12. One-Fund Separation à la Ross.

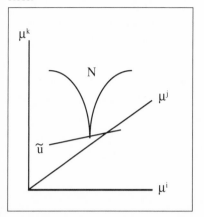

Figure 13. The First Order Conditions in Two Dimensions.

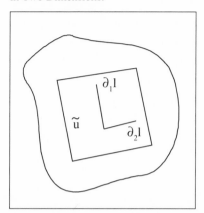

optimal path is 1. Again we are on the boundary of P in mixture space. Figure 14 shows the picture. Obviously rotating $\tilde{u}$ in the appropriate direction leaves the maximum unchanged. Since the dimension of all one-forms on a two-dimensional space is 2, the number of coordinates of the forms, and since the dimension of a line in 1, it is to be expected that to squeeze in the extra dimension, the tangent vector spaces cannot be of full "rank".

Figure 14. Two-Fund Separation à la Ross.

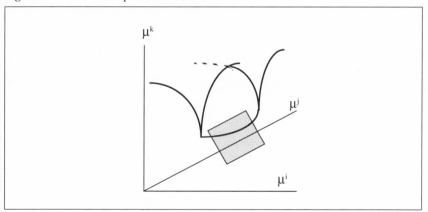

Case d: C-S 2-Fund Separation   As we have seen, changes in wealth will in general rotate the one form in moment space. When we maximise on P, this rotation will be along the surface of P. If this is to produce a line in asset space, the rotation of the one forms must be exactly offset by the curvature of the portfolio manifold along the line.

A change in W, then generates a field of forms. If we associate each member of this field with a point on the optimal line with the rule $\tilde{u}(W)$ chooses $\theta(W)$, then we have a picture like Figure 15. These forms may be integrated to form a smooth "cap." If we have C-S 2 Fund Separation, the cap fits the manifold P along the required path, Figure 16.

Figure 15. The One-Form Cap.

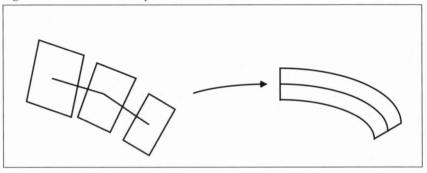

This will never happen on any manifold which can be "dented" i.e. whose curvature can be changed at some points but not others. Presumably the fact that manifolds can be dented explains why in general this separation is rare.

All of the cases of dimension reduction given above are subsumed in the following idea. Static portfolio theory is a quadruple

$$P, g, \tilde{u}, \Gamma$$

where

1. P is a manifold

2. g is a metric

3. $\tilde{u}$ is a closed one-form on P

4. Γ is the mixture connection.

The dimension reduction theorems are all associated with special shapes of P in M. It seems reasonable to suppose that these shapes are symmetries representable by Lie groups, but the precise description of these groups has not yet been developed.

Figure 16. Two-Fund Separation à la Cass-Stiglitz.

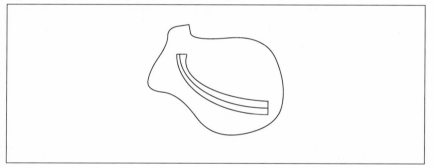

5. DIVERSIFICATION AND GENERALISED PREFERENCES

None of the previous section requires that the individual maximise expected utility. This is because all the results take place in the tangent space. However, some results in investor behavior do depend on expected utility maximisation. To indicate how portfolio theory will change as we move to generalised preferences, we consider the problem of diversification.

It is well known, see Machina [1982] that for a risk averse expected utility maximiser who is indifferent among n assets X_i the internal portfolio $P(\theta) = \theta^i X_i, 0 < \theta^i < 1$ is always as least as good as X_i. With General Expected Utility, however, this is not the case, see Dekel [1987]. We have recently given the necessary and sufficient conditions for diversification in this case, and in Figure 17, we sketch the problem for two assets.

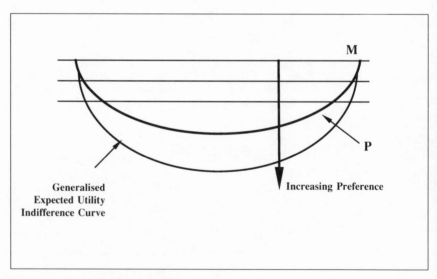

Figure 17. Diversification with General Preferences.

We embed the two assets in moment space as before so that indifference curves for expected utility maximisers are flats. As Roell [1985] has shown, portfolios $f^p(\theta)$ are mean preserving spreads of mixtures $f(\theta)$. Thus the direction of increasing preference is as indicated, and thus it pays to diversify. With generalised preferences, however, the indifference curves are no longer flats. Thus, as illustrated, diversification may not pay. The problem is that with General Expected Utility we must contend with two curvatures, the curvature of the portfolio set P and the curvature of the field of $\tilde{u}$. The interaction of these curvatures determines whether or not it pays to diversify, see Russell [1988a].

6. CONTINUOUS TIME PORTFOLIO THEORY

In this section we shall apply geometric methods to the study of continuous time portfolio theory. Financial assets traded in continuous time will be treated as a **Distribution** (in the Schwartz/Sobolev sense) with **Support** the **Origin**, ("DSO"). The moments of such assets can be easily calculated using the rules of distribution theory.

Secondary assets, i.e., assets such as stock options and warrants whose value is determined by the value of some primary asset, are shown to also be DSOs if the primary asset is. Their moments and the cross moments of the primary and secondary asset are also easily calculated using the rules of distribution theory. This is shown to permit the derivation of dynamic riskless hedges both exactly for continuous path Markov process, and approximately when the process involves jumps.

The Stochastic Dynamics of Asset Prices

To see the relationship between distribution theory and portfolio theory clearly, assume that the stochastic dynamics of the price of the primary asset y is given by a Log Wiener process. It is customary to represent such a process by the Stochastic Differential Equation (S.D.E.) which describes the behavior of the sample paths,

$$dy = \mu y(t)\,dt + \sigma y(t)\,dW \tag{1}$$

where

$$y(t) \text{ is the price of the stock at time } t$$

and

$$W(t) \text{ is a Wiener process.}$$

However, as Itô has shown, every S.D.E. is equivalent to some Partial Differential Equation which describes the evolution not of the sample paths, but of the probability density function generating the sample paths.

For the case of equation (1) the relevant equation, (Fokker-Planck equation) is

$$\frac{\partial p}{\partial t}\,(y, t \mid y_0, 0) = -\frac{\partial}{\partial y}\,(\mu y(t))\,p\,(y, t \mid y_0, 0) + \frac{1}{2}\,\frac{\partial^2}{\partial y^2}\,[(\sigma y(t))^2\,p(y, t \mid y_0, 0)] \tag{2}$$

where $p(y, t \mid y_0, 0)$ is the probability that the price will be y at time t given that it was y_0 at time 0.

Since at time $t = 0$ the price of the stock is y_0 with probability $p = 1$, the initial condition for this partial differential equation is

$$p(y, 0 \mid y_0, 0) = \delta(y - y_0) \tag{3}$$

where $\delta(\ )$ is the Dirac delta function. If we now explicitly evaluate equation (2) we have

$$\frac{\partial p}{\partial t} (y, 0, | \, y_0, 0) = (\sigma^2 - \mu) \, \delta(y - y_0) \ +$$

$$(2\sigma^2 - \mu) \, y \delta^1 (y - y_0) \ + \qquad (4)$$

$$\frac{1}{2} \, (\sigma y)^2 \, \delta^2 (y - y_0)$$

Finally setting $x = y - y_0$ so that x is the change in price of the stock, we must have

$$p(x, 0, | \, y_0, 0) = \ p(y_0 + x, 0, | \, y_0, 0)$$

from which we see that

$$p(x, dt, | \, y_0, 0) \ = \ \delta + [c_0 \delta + c_1 \delta^1 (x) + c_2 \delta^2 (x)] \, dt \qquad (5)$$

where

$$c_0 = \sigma^2 - \mu$$
$$c_1 = (2 \, \sigma^2 - \mu) \, y_0$$
$$c_2 = (1/2) \, (\sigma y_0)^2.$$

This equation states that the transitional probability density for the change in price of the asset is a linear combination of the delta function and its first and second derivatives. Such an object is certainly a distribution, and since its support is the origin, it is indeed a DSO. Because the transitional density is a distribution and not a function, no meaning can be given to its point values. The difficulties which arise when one does try to give point values meaning is known in financial economics as the put option paradox, Grinblatt and Johnson [1987]. There is no paradox when the true nature of delta functions is recognised Russell [1988b].

Assets as DSOs

We have seen that we can associate a DSO with the Log-Wiener process. Can we associate a DSO with every continuous Markov process? The answer to this question is a qualified yes. The following facts then provide a new foundation for the theory.

Fact One. A distribution whose support is the origin is a *finite* combination of derivatives of the Dirac delta function. The highest order of derivative plus one is called the order of the distribution. See Choquet et al. [1977] p. 363.

Fact Two. Every continuous time continuous path Markov process can be represented exactly as a DSO of order 3. See Pawula [1967]. Every continuous time jump Markov process can be approximated by a DSO where the approximation is given by the Kramers-Moyal expansion (Russell [1988d]).

Fact Three. If $g(x)$ is differentiable of order j at the origin,

$$\int_a^b g(x)\delta_j(x)\,d(x) = \begin{cases} 0, & \text{for } 0 \notin [a, b] \\ (-1)^j D^j g(0), & \text{for } 0 \in [a, b\}. \end{cases}$$

where D^j is the differential operator of order j. See Choquet et al. [1977] p. 360.

7. CONTINUOUS TIME PORTFOLIOS AND SECONDARY ASSETS

Let S_t be a continuously traded asset whose price change when held for time t, $x(t)$, has a generalised probability density function which satisfies

$$p(x, 0, \mid y_0, 0) = \delta(x)$$

$$\frac{\partial p}{\partial t}(x, 0, \mid y_0, 0) = \sum_{i=0}^{K-1} c_i \delta^i(x)$$

where y_0 is the initial price of the asset, and d^i is the ith derivative of d. Clearly $\partial p / \partial t$ is a DSO of order k.

The moments μ^i of x associated with holding S_t for the length of time dt are given by

$$\mu^i(x, dt) = \mu^i(x, 0) + \frac{\partial \mu^i}{\partial t}(x, 0)dt$$

$$= \frac{\partial \mu^i}{\partial t} dt$$

$$= (-1)^i \, i \, ! \, c_i \qquad (i \leq k)$$

$$= 0 \qquad \text{otherwise}$$

Under these assumptions on S_t, the sequence of moments of a continuously traded asset will always be of *finite* length. Recently there has been a great deal of interest in the valuation of secondary assets, i.e., assets whose value is given as a function of the value of some primary asset at or before some time. If the primary asset can be associated with a DSO, what about secondary asset? In particular, what are its moments, and what are the cross moments?

We have

Theorem One. *Let the change in price x of an asset S_t have a generalised density as above. Let O_t be a derivative asset. That is $z(t)$, the change in price of O_t is given by a mapping say $f: R \times T \rightarrow R$ so that $z(t) = f(x(t), t)$, with $f(0, 0) = 0$. O_t is then a continuously traded asset whose value is derived from the value of S_t. Then*

1. z_t has a density which is a DSO of order k.
2. $\mu^i(z, dt)$, the ith moment of z_t is given by

$$\mu^i = \sum_{m=1}^{k-1} (-1)^m c_m D^m (f^i(0, 0)) \, dt$$

where f^i is the ith power of f.

3. $\mu^{ij}(xz, dt)$, the ijth cross moment of x_t and z_t, is given by

$$\mu^{ij}(xz, dt) = 0 \quad i > k - 1$$

$$\mu^{ij}(xz, dt) = c_i(-1)^i\, i\, !\, f^j(0, 0)$$

$$+ \Sigma(-1)^m \frac{m!}{(m-i)!}\, c_m D^{m-i} f^j(0, 0) \quad \text{otherwise}$$

In particular

$$\text{COV}(xz, dt) = (-c_1 f(0) + 2c_2 D^1 f(0, 0) - 3c_3 D^2 f(0, 0) + \ldots)dt$$

4. There exists a portfolio P of S_t and one derivative asset, say O, such that

$$\mu^i(P) = 0, \qquad i > 1$$

Proof. (1), (2), and (3) are applications of Fact Three. (4) follows from the application of (2) and (3) to the portfolio $S_t - (1/f_1)\, O_t$.

$$\text{Q.E.D.}$$

For example, when the price behavior of S_t is given by (2), it is easy to see that the first two moments of the stock price change are

$$\mu(x, dt) = -c_1\, dt$$

$$\mu^2(x, dt) = 2c_2\, dt.$$

Suppose that the value of the secondary asset O_t in terms of the stock is written as some function $f(y, t)$, and that we hold $\alpha(t)$ of this asset at time t.

Then the change in the value of the secondary asset position, which we write as z, can be written as

$$z = [\alpha f](y, dt) - [a f](y_0, 0) \tag{6}$$

Expanding in Taylor series around y_0 gives

$$z = \alpha(0)\, (f_1(\ )x + \frac{1}{2}\, f_{11}(\ )\, x^2 + \ldots) +$$

$$\frac{da}{dt}\, (0)f(\ )dt + \alpha(0)f_2(\ )dt + \ldots \tag{7}$$

where all derivatives are evaluated at $(y_0, 0)$.

Notice that expansion of the Taylor series in terms of t higher than dt yields terms of second order. We could expand in terms higher than $x^{\wedge 2}$ but this is pointless since by Fact One,

$$\int x^n\, \delta^m = 0 \qquad n > m \tag{8}$$

Now we may calculate the relevant moments of z and the cross-moments of x and z. We have,

$$\mu\,(z, dt) = [-\alpha\,(f_1 c_1 + f_{11}\, c_2 + f_2) + \frac{d\alpha}{dt} f]dt$$

$$\mu^2(z, dt) = [(\alpha f_1)^2\, 2\, c_2]dt \tag{9}$$

$$\mu\,(xz, dt) = [\alpha f_1\, 2\, c_2]dt$$

where all variables are evaluated at $(y_0, 0)$. All higher moments are zero. By setting $\alpha = -1/f_1$ we see that we may create a portfolio of the stock and the option which has no variance. The expected change in price of this hedge portfolio is $\mu\,(x, dt) + \mu\,(z, dt)$ which is easily seen to be

$$[f_{11}c_2 + f_2 + \frac{da}{dt}\, f\,]\, dt. \tag{10}$$

The last term in this expression is the change in the holding of the secondary asset times that asset's initial price. Subtracting this from (10) gives the riskless price change net of additions in funds to the portfolio. This expression implies a riskless return per dollar invested in this portfolio of stock and secondary asset, and this return must be equated to the instantaneous interest factor $r\ dt$ to rule out arbitrage. This in turn implies a partial differential equation which can be solved for the secondary asset's price.

Theorem One generalises the famous Black-Scholes [1973] hedging argument to any stochastic process which can be represented by a DSO. Since from Fact Two every Markov process can be approximated by a DSO, the theorem states that up to the approximation provided by the Kramers-Moyal [1940, 1949] expansion every Markov process can be hedged with one secondary asset. Thus Distribution theory provides a basis for continuous time portfolio theory in much the same way that random variable theory provided a basis for discrete time portfolio theory. In an important sense, however, continuous time portfolio theory is simpler than discrete time portfolio theory. If we think of the space of portfolios as having as coordinates the moments of the portfolios, continuous time portfolios are typically embedded in a low dimension space.

The reason for this is that continuous time portfolio theory deals with assets in the tangent space to the space of random variables, and for the diffusion processes of finance, this tangent space is more symmetric than the underlying asset space. This symmetry allows us to ignore most of the moment coordinates thus allowing hedging with a small number of assets.

8. CONCLUSION

The purpose of this paper has been to indicate the power of geometric methods as a unifying device in the study of individual behavior under uncertainty. We have seen that a rich theory can be built on the differential geometry of families of probability density functions, and one-form fields defined on them. Since the results of experimentalists seem to suggest that the axioms of expected utility maximisation are frequently and systematically violated, the value of such a framework seems obvious.

NOTES

1. Strictly speaking a one form provides more than we need if our interest is only in equilibrium choice. Multiplication of the field $\tilde{u}$ by a positive constant a would have no effect on the sign of $\tilde{u}(\theta)\,\bar{v}(\theta)$ and therefore could not affect behavior governed by Hypothesis One. Multiplying the field by a positive constant is equivalent to performing an increasing monotonic transformation on the functional which generates the field. The equivalence class $a\tilde{u}$ is called a hypersurface contact element and this is the fundamental geometric object which underlies non-stochastic as well as stochastic choice.

REFERENCES

Allais, M. (1952). Fondements d'une theorie positive des choix. *Econometrie, 40,* 257-332.

Amari, S-i. (1984). *Differential-geometric methods in statistics.* Berlin: Springer-Verlag.

Atkinson, C. & Mitchell, A. F. (1981). Rao's distance measure. *Sankya, 43A,* 345-365.

Barnsdorf, N. O. (1978). *Information and exponential families in statistical theory.* New York: Wiley.

Bell, D. E. (1982). Regret in decision making under uncertainty. *Operations Research, 30,* 961-981.

Black, F. & Scholes, M. (1973). The pricing of options and the evaluation of corporate liabilities. *Journal of Political Economy, 81,* 637-653.

Becker, J. & Sarin, R. (1988). Lottery dependent utility. University of California at Los Angeles working paper.

Camerer, C. (1988). An experimental test of several generalised utility theories. *Journal of Risk and Uncertainty,* forthcoming.

Cass, D. & Stiglitz, J. (1970). The structure of investor preferences and asset return separability. *Journal of Economic Theory, 10,* 122-160.

Centsov, N. N. (1972). *Statistical decision rules and optimal inference.* Rhode Island: A.M.S.

Chew, S. H. (1983). A generalization of the quasi-linear mean, etc. *Econometrica, 51,* 1065-1092.

Chew, S. H. & Macrimmon, K. (1979). Alpha-nu utility theory, etc. University of British Columbia working paper No. 686.

Choquet-Bruhat, Y., Dewitt-Morette, C. & Dallard-Bleik, M. (1977). *Analysis, manifolds and physics.* Amsterdam: North-Holland.

Dawid, A. F. (1975). Discussion to Efron's paper. *Annual Statistics, 3,* 1231-1234.

Debreu, G. (1959). *Theory of value.* New Haven: Yale University Press.

Dekel, E. (1987). Asset demands without the independence axiom. University of California at Berkeley mimeograph 1-10.

Efron, B. (1975). Defining the curvature of a statistical problem. *Annual Statistics, 3,* 1189-1242.

Grinblatt, M. & Johnson, H. (1988). A put option paradox. *Journal of Financial and Quantitative Analysis.*

Fishburn, P. C. (1984). SSB utility theory, etc. *Mathematical Social Sciences, 8,* 253-285.

Gardiner, C. W. (1985). *Handbook of stochastic methods.* New York: Springer-Verlag.

Kahneman, D. & Twersky, A. (1979). Pospect theory. *Econometrica, 47,* 263-291.

Kramers, H. A. (1940). *Physica, 7,* 284-285.

Loomes, G. & Sugden, R. (1986). Disappointment and dynamic choice under uncertainty. *Review of Economic Studies, 53,* 271-282.

Loomes, G. & Sugden, R. (1982). Regret theory. *Economic Journal, 92,* 805-824.

Lighthill, M. J. (1978). *Fourier analysis and generalised functions.* Cambridge: Cambridge University Press.

Machina, M. (1984). The economic theory of individual behavior towards risk, etc. Working paper.

Machina. M. (1982). Expected utility analysis without the independence axiom. *Econometrica, 50,* 277-323.

Marschak, J. (1950). Rational behavior, uncertain prospects, and measurable utility. *Econometrica, 18,* 111-141.

Markowitz, H. (1959). *Portfolio selection.* New Haven: Yale University Press.

Moyal, J. E. (1949). Journal of the Royal Statistical Society, 11, 151-171.

Pawula, R. F. (1967). Approximation of the linear Boltzmann equation by the Fokker-Planck equation. *Physical Review, 162,* 186-188.

Pfanzagl, J. (1982). *Contributions of a general asymptotic statistical theory.* New York: Springer-Verlag.

Quiggin, J. (1982). A theory of anticipated utility. *Journal of Economic Behavior and Organisations, 3,* 323-343.

Rao, C. R. (1945). Information and accuracy. *Bulletin of the Calcutta Mathematical Society, 37,* 81-91.

Roell, A. (1985). Risk aversion in Yaari's model. L.S.E. mimeograph No. 1-16.

Ross, S. (1978). Mutual fund separation in financial theory. *Journal of Economic Theory, 17,* 254-286.

Russell, T. (1988A). Risk aversion and asset diversification for general preferences. Santa Clara University mimeograph.

Russell, T. (1988B). Schwartz distributions resolve the put paradox. Santa Clara University mimeograph.

Russell, T. (1988C). Tricks with Engel curves. Santa Clara University mimeograph.

Russell, T. (1988D). Continuous time portfolio theory and the Schwartz-Sobolev theory of distributions. *Operations Research Letters, 7,* 159-162.

Samuelson, P. A. (1965). Rational theory of warrant pricing. *Industrial Management Review, 6,* 13-69.

Samuelson, P. A. (1983). *Foundations of economic analysis: Enlarged edition.* Cambridge, Massachusetts: Harvard University Press.

Sato, R. (1981). *Theory of technical change and economic invariance.* New York: Academic Press.

Sato, R. & Ramachandran, R. (1989). Symmetry and conservation: An introduction. *Symmetry and conservation laws: With applications to economics and finance*. Boston: Klewer Academic Press.

Spivak, M. (1977). *A comprehensive introduction to differential geometry*. Boston: Publish or Perish Press.

Weber, M. & Camerer, C. (1987). Recent developments in modelling preferences under risk. *O. R. Spectrum, 9*, 129-151.

Yaari. (1987). The dual theory of choice under risk. *Econometrica, 55*, 95-115.

CHAPTER 8

Symmetries, Dynamic Equilibria, and the Value Function

John H. Boyd III

1. INTRODUCTION

This paper presents a geometric approach (symmetries) to dynamic economic problems that integrates the solution procedure with the economics of the problem. Techniques for using symmetries are developed in the contexts of portfolio choice, optimal growth, and dynamic equilibria. Information on preferences, budget sets, and technology is combined to explicitly compute the solution. By focusing on the geometry of the underlying economic structure, the symmetry method can handle many types of problems with equal ease. Given an appropriate economic structure, it is immaterial whether the problem is in continuous or discrete time, is deterministic or stochastic with a Brownian, Poisson or other process, uses a finite or infinite time horizon, or even whether the rate of time preference is fixed or variable. These details are unimportant as long as the geometry is unchanged. All cases are treated in a unified manner.

*I would like to thank Fwu-Ranq Chang, whose encouragement was invaluable, and who, with Robert Becker, saw this paper through various revisions. I would also like to thank Rabah Amir, Rolf Fare, Roy Gardner, Robert King, C. Knox Lovell and Paul Romer for their comments and suggestions.

Symmetries exhibit various interesting features. A major strength of the symmetry technique is its ability to ferret out the solutions to complex models with simple underlying economic structures. A previously unsolved optimal growth model with both time-varying discount rates and technology is easily solved via symmetries. Symmetries can also be used to transform problems into a simpler form, as will be demonstrated for hyperbolic absolute risk aversion (HARA) felicity. Symmetries are well-adapted for dealing with equilibrium problems. For example, they can demonstrate the uniqueness of equilibrium. Finally, the symmetry method does not require the use of the transversality condition, although it will hold if necessary for an optimum.

In contrast, most dynamic economic models are solved on a case-by-case basis. Although the methods of dynamic programming bring some order to the subject, they do not fully exploit the economic structure of the models. Stochastic control problems are a case in point. They can rarely be explicitly solved even though the Bellman partial differential equation characterizes the solutions.

Explicit solutions have been found in a few cases involving portfolio selection and asset pricing.[1] Although Samuelson (1969) derived the solution for one problem, most of these solutions were obtained by the method of "divine revelation."[2] Guess the solution and plug it in. If it works, fine; if not, try another guess. This reliance on trial solutions is not totally satisfactory. The most general result was found by Danthine and Donaldson (1981).[3] Their theorem still leaves the solution technique unconnected with the economic structure of the problem. Why the problem has a solution of this form is left unexplained.

Symmetries explain the form of the solution in terms of the geometry of preferences and technology. Mathematically, these geometric relationships are expressed by using certain transformations (symmetries) that are based on the economic structure. The symmetries generate generalized notions of homogeneity and homotheticity, such as Sato's (1981) holotheticity. This generalized homotheticity turns symmetries into powerful tools. With them, we can discover many of the economically important properties of the solutions. Their full power is most apparent when dealing with more complex problems. This is especially true in dynamic competitive equilibrium problems. The effects on many agents can easily be aggregated, even when the agents are heterogeneous. Section Two gives some elementary examples of solution via symmetries. Section Three contains mathematical preliminaries on stochastic models, sets up the basic abstract framework, and presents a theorem relating symmetries to the value function.

The remainder of the paper presents a number of examples of symmetries in action. Section Four uses a simple linear symmetry to examine various portfolio models previously studied by Merton (1969, 1971, 1973), Fischer (1975) and Gertler and Grinols (1982). Two main conditions are used. The first is linear homogeneity of the state equation, as in typical budget constraints. The second condition is that the felicity function be either homogeneous or logarithmic. The constant relative risk aversion felicity functions obey this restriction. The last part of Section Four shows how the symmetry approach works in more complex cases where the felicity function is neither homogeneous nor logarithmic. Symmetries apply to Merton's portfolio problem even for the more general hyperbolic absolute risk aversion felicity functions. The same arguments apply to more general stochastic processes. In particular, they apply for Poisson processes.

Section Five examines Ramsey problems with Cobb-Douglas production functions. Two types of model are considered. In the first, the symmetries determine the form of the value function for a stochastic Ramsey problem with discounting originally solved by Mirman and Zilcha (1975). The undiscounted version (Mirman and Zilcha, 1977) can be handled in exactly the same way. It makes no difference to the symmetries if the problem is discounted. The second example is a variant of the Mirman and Zilcha model inspired by Mitra (1979) that allows for non-stationary production and discount rates. Many time-varying models may be explicitly solved via the symmetry technique.

Section Six examines a sample equilibrium problem. This is Michener's (1982) version of Lucas' (1978) asset pricing model. In fact, the symmetry applies to the more general asset pricing model of Brock (1982). Even in this wider framework, they can show that Lucas' stationary asset pricing function gives all possible equilibrium prices. Interestingly, the transversality condition need not be invoked to prove uniqueness of the price sequence.

2. ELEMENTARY EXAMPLES OF SYMMETRIES

Before delving into the technical details, I want to examine some simple problems where symmetry properties apply. These are standard economic models where geometric reasoning about the shape of indifference curves and feasible sets yields useful information. Symmetries were originally developed by physicists and crystallographers to formalize notions of shape via invariance concepts.[4] For example, when a sphere is rotated, its shape is unchanged. One way of expressing the fact that a sphere is round is to note that all rotations map the sphere onto itself. The symmetric shape of the sphere is described by its invariance under rotation. Equivalently, rotation is a symmetry of the sphere. When a crystal is rotated through certain angles, its shape is also the same. These rotations, and no others, are symmetries of the crystal. Similarly, many physical laws are unchanged under translation or rotation. Translation and rotation are symmetries of these physical laws. Typically, these symmetries form a group—the composition of any two symmetries is a symmetry, and the inverse of any symmetry is a symmetry.

One of the simplest models in economics is consumer choice in a two commodity world. The consumer faces fixed prices p_x and p_y for

Figure 1.

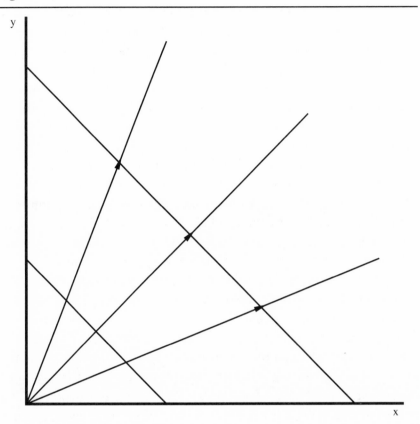

goods x and y. Consider the isocost lines. When $\lambda > 0$, the transformation $T_\lambda(x, y) = (\lambda x, \lambda y)$ maps each isocost into another isocost (with cost multiplied by λ). The mapping of a representative isocost and the expansion lines are shown in Figure 1. The uniform expansion T_λ maps the set of isocosts onto itself. Any T_λ with $\lambda > 0$ is a symmetry of isocosts. Note that T_λ's form is a group since $T_\lambda T_\mu = T_{\lambda\mu}$ for λ, μ > 0 and T_1 is the identity map.

Similar arguments apply to budget sets, where the budget set with income m is mapped onto the budget with income λm. The symmetry expresses the fact that the shape of the budget set does not depend on income. Income only changes its size. Similarly, the fact that only relative prices matter expresses a symmetry under scaling of prices.

This has special significance when preferences are homothetic. With homothetic preferences, $(x, y) \geq (w, z)$ if and only if $(\lambda x, \lambda y) \geq (\lambda w, \lambda z)$. The transformation T_λ leaves the collection of indifference curves (the indifference map) unchanged for homothetic preferences. The symmetry is not only a symmetry for budget sets, it is a symmetry for indifference curves too. Now suppose that income increases from m to m'. The budget set expands uniformly, as in Figure 1, which also illustrates the expansion paths. This expansion corresponds to the case $\lambda = m'/m$. When preferences are homothetic, this uniform expansion leaves the indifference map unchanged. The optimum in the original budget set is mapped to the new optimum. At the optimum, the relative quantities of each good are the same, only the absolute amount changes. The income expansion paths must be straight lines through the origin. When utility is homogeneous, plugging this in the utility function reveals that the indirect utility function is homogeneous in income.

When there are many goods, or a more complicated feasible set, the geometry may not look quite so simple. Rather than inspecting the shapes of budget sets and indifference maps on a diagram, we must use more powerful analytic methods. The uniform expansion used to define homogeneity must be replaced by a more complex symmetry. The properties of these symmetries may yield similar results in the end.

One such case arises when there is a production possibility set, rather than a budget set. Increasing the endowment can cause a lopsided expansion (nonisotropic dilation) of the feasible set. Provided the indifference map is invariant, the expansion takes optima to optima. We can then easily calculate the indirect utility (value) function. For example, consider the feasible set defined by $y = (m - x)^{1/2}$ (this could arise from a two-period capital accumulation model with production function $f(k) = \sqrt{k}$). As m is increased to λm, the

Figure 2.

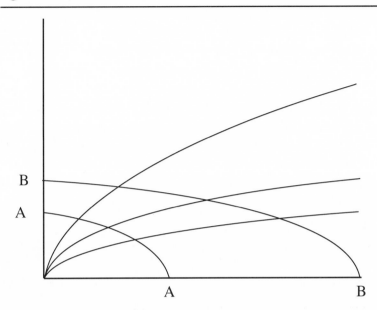

production frontier expands in a lop-sided manner. This expansion is performed by the symmetry $T(x, y) = (\lambda x, \lambda^{1/2}y)$. Parabola AA in Figure 2 has $m = 4$ while parabola BB has $m = 16$. The other parabolas are the expansion paths running through $(1, 3)$, $(1, 1)$ and $(3, 1)$.

Consider the point $S = (2, \sqrt{2})$ in Figure 3. When m rises from 4 to 16, this is mapped to $T = (8, 2\sqrt{2})$ by the symmetry ($\lambda = 4$). The indifference maps of most constant elasticity of substitution (CES) utility functions are distorted by this transformation . For example, the Leontieff utility function $u_1(x, y) = \min (x, y\sqrt{2})$ is transformed into $u_2(x, y) = \min (\lambda x, (2\lambda)^{1/2}y)$. The indifference curves are not invariant under the production symmetry. Although S is optimal for u_1 when $m = 4$, the optimum for $m = 16$ lies at R. Among the CES utility functions, only the Cobb-Douglas indifference curves are undistorted, and so only Cobb-Douglas utility has T as the new optimum when S is the old optimum. (In fact, only one Cobb-Douglas utility function,

Figure 3.

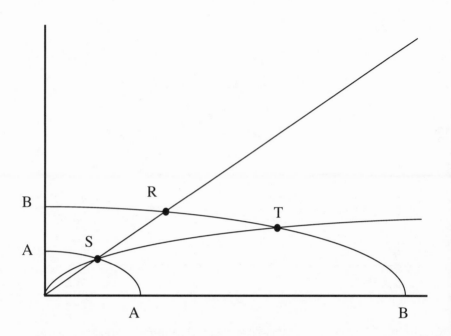

$u(x, y) = x^{1/3} y^{2/3}$, has S as the original optimum.) For general CES utility, the actual location of the new optimum depends on the elasticity of substitution σ. When S is the original optimum, the new optimum will be R when $\sigma = 0$ (Leontieff), between R and T for σ between 0 and 1, T for $\sigma = 1$ (Cobb-Douglas), and to the right of T for σ greater than 1. In finite-horizon discounted models, homogeneous felicity yields a CES utility function, while logarithmic felicity gives Cobb-Douglas utility. As in the two-period model, only the logarithmic case is well-behaved.

Although CES utility is generally ill-behaved under this symmetry, there is a related class of utility functions which are well-behaved. Consider $u(x, y) = [\delta x^{-\rho} + (1 - \delta)y^{-2\rho}]^{-v/\rho}$ for $v > 0$, $\rho > -1$ and $\rho \neq 0$. Clearly $u(\lambda x, \lambda^{1/2}y) = \lambda^v u(x, y)$, so indifference curves are mapped to indifference curves. For any of these utility functions, optima will be mapped to optima. In fact, a nonisotropic dilation $S(x, y) = (\lambda^\alpha x, \lambda^\beta y)$ will be a symmetry for any utility function of the form $u(x, y) = v(x^{1/\alpha}, y^{1/\beta})$ where v is any CES utility function since $u(\lambda^\alpha x, \lambda^\beta y) = v(\lambda x^{1/\alpha}, \lambda y^{1/\beta})$. The indifference map of this modified CES function is invariant under the nonisotropic dilation S.

3. SYMMETRIES IN MARKOV DECISION MODELS

A form of Markov decision model provides a natural setting for the investigation of symmetries in dynamic models. It provides enough generality to discuss most known interesting examples, while remaining simple enough to easily work with. However, as symmetries can occur in many other types of models, the results here are more illustrative than definitive. The main theorem is not intended to apply to every possible case.

3.1 Stochastic Preliminaries

We start by recalling some basic probability theory. Consider a probability space $(\Omega, \mathfrak{F}, P)$ where $\mathfrak{F}$ is a σ–field on the space Ω and P is a probability measure on the measurable space $(\Omega, \mathfrak{F})$. A *stochastic process* on Ω is a collection of random variables ($\mathfrak{F}$-measurable functions) $\{X_t : t \in \mathfrak{T}\}$. We will restrict our attention to processes indexed by either $\mathfrak{T} = [0, T]$ (for continuous processes) or the set $\mathfrak{T} = \{0, ..., T\}$ (for discrete processes). Of course, T can be infinite with $\mathfrak{T} = [0, \infty)$ or $T = \{0, ...\}$.

Let $\mathfrak{G}$ be a σ-field with $\mathfrak{G} \subset \mathfrak{F}$ and Y a real-valued random variable on Ω. The *conditional expectation* $\mathcal{E}[Y \mid \mathfrak{G}]$ of Y is the function such that $\int_B Y \, dP = \int_B \mathcal{E}[Y \mid \mathfrak{G}] \, dP$ for all $B \in \mathfrak{G}$. Such a function exists and is almost surely unique by the Radon-Nikodym theorem. When $\mathfrak{G} =$

$\sigma[X]$, the σ-field generated by sets of the form $\{\omega: X(\omega) = \alpha\}$, we write $\mathcal{E}[Y \mid X]$. Let $\mathcal{X}_t$ be the σ-field generated by all sets of the form $\{\omega: X_s(\omega) = \alpha\}$ for $s \leq t$. A process $\{X_t\}$ is *Markov* if $\mathcal{E}[Y \mid \mathcal{X}_t] = \mathcal{E}[Y \mid X_t]$ for all $\mathfrak{F}$-measurable function Y. This says that the current state of the process X_t summarizes all of the information contained in its history.

When time is continuous, the m-dimensional Wiener process Z (normalized Brownian motion) is an example of a Markov process. Recall that Z is a Wiener process if:

  (1) $Z_0 = 0$ a.s.,
  (2) the increments $Z_{t_i} - Z_{t_{i-1}}$ are independent when $0 \leq t_0 \leq \ldots \leq t_n$,
  (3) for all $s \leq t$, $Z_s - Z_t$ is normally distributed with mean 0 and covariance matrix $(s - t) \times I_m$, and
  (4) $t \mapsto Z_t(\omega)$ is a continuous function for each $\omega \in \Omega$.

Given a Wiener process Z, a family of σ-fields $\mathfrak{F}_t$ is non-anticipating with respect to Z if:

  (1) $\mathfrak{F}_s \subset \mathfrak{F}_t$ for all $s \leq t$,
  (2) $\mathfrak{F}_t \supset \sigma[Z_s]$ for all $s \leq t$, and
  (3) $\mathfrak{F}_t$ is independent of $\sigma[Z_u - Z_t]$ for all $u \geq t$.

A measurable function $\sigma: [0, T] \times \Omega \to \mathbb{R}$ is *non-anticipating* with respect to the family $\mathfrak{F}_t$ if $\omega \mapsto \sigma(t, \omega)$ is $\mathfrak{F}_t$-measurable for all $t \in [0, T]$ and the integral $\int_0^T \mid \sigma(t, \omega) \mid^2 dt$ is finite with probability one.

Let $0 = t_0 < t_1 < \ldots < t_n = T$ be a partition of $[0, T]$, and suppose that σ is a non-anticipating step function that is constant on each interval $[t_i, t_{i+1})$. Define the *Itô integral* of σ by $\int_0^T \sigma(t, \omega) \, dZ = \Sigma_{i=1}^n \sigma(t_{i-1}, \omega) [Z_{t_i}(\omega) - Z_{t_{i-1}}(\omega)]$. Now consider an arbitrary non-anticipating function σ. There is a sequence of non-anticipating step functions σ_n which converges in mean square to σ. There is a sequence of non-anticipating step functions σ_n which converges in mean square to σ. The Itô integrals of the σ_n also converge in mean square to a function denoted by $\int_0^T \sigma \, dZ$. This function does not depend on the particular approximating sequence used, and is the Itô integral of σ.

By varying the terminal time, we obtain a stochastic process $X_t = \int_0^T \sigma \, dZ$. This relation is often expressed by the stochastic differential equation $dX = \sigma\, dZ$. An *Itô process* is a process X with stochastic differential $dX = \alpha(X, t)\, dt + \sigma(X, t)\, dZ$ where α is an $n \times 1$ matrix, σ is $n \times m$ matrix and Z is an m-dimensional Wiener process. Any Itô process is Markov, although not all Markov processes are Itô. A fundamental result concerning Itô processes is Itô's Lemma.

Itô's Lemma.

Let $u(t, x) : [0, T] \times \mathbb{R}^n \to \mathbb{R}$ be a continuously differentiable function. Suppose that $X(t)$ is an n-dimensional Itô process with stochastic differential $dX = \alpha(t)\, dt + \sigma(t)\, dZ$ where Z is an m-dimensional Wiener processes. Let $Y(t) = u(X(t), t)$. Then the stochastic differential of y exists on $[0, T]$ and is given by $dY = [\partial u/\partial t + (\partial u/\partial x)\alpha + (1/2)\, \mathrm{tr}(\sigma\sigma'\partial^2 u/\partial x^2)]\, dt + (\partial u/\partial x)\sigma\, dZ$.

When the processes are all one-dimensional, this says $dY = [\partial u/\partial t + \alpha\, \partial u/\partial x + (\sigma^2/2)\, \partial^2 u/\partial x^2]\, dt + (\partial u/\partial x)\sigma\, dZ$. For example, if $u(x) = x^2$ and $dX = dZ$, Itô's Lemma shows that $d(Z^2) = dt + 2Z\, dZ$. In the two-dimensional case with $dX_1 = \alpha_1\, dt + \sigma_1\, dZ$ and $dX_2 \times \alpha_2\, dt + \sigma_2\, dZ$, Itô's Lemma yields $d(X_1 X_2) = [\alpha_1 X_2 + \alpha_2 X_1 + \sigma_1\sigma_2]\, dt + (\sigma_1 X_2 + \sigma_2 X_1)\, dZ$.

Alternatively, the stochastic differential can be obtained by using Taylor's formula to derive the differential. However, some second order terms must be retained. For stochastic differentials, the convention is that $(dt)^2$, $dt\, dZ$ and $dZ\, dt$ are 0 while $(dZ)^2 = dt$, even for m-dimensional processes. Using this we quickly obtain the product and quotient rules $d(X_1 X_2) = X_1\, dX_2 + X_2\, dX_1 + dX_1\, dX_2$, and $d(X_1/X_2) = [X_2\, dX_1 - X_1\, dX_2]/X_2^2 + X_1\, dX_2^2/X_2^2 - dX_1\, dX_2/X_2^3$. These rules will prove useful in Section Four. Arnold (1974) and Malliaris and Brock (1982) contain more detailed information on Itô integration and Itô processes.

We will assume the Markov process obeys a system of stochastic differential (or difference) equations, $L(X) = 0$. These are *evolution equations*. For example, suppose time is discrete and capital x_t

follows a logarithmic Brownian motion $\log x_{t+1} = \log x_t + \varepsilon_t$ where the ε_t are identically and independently distributed. The evolution equation is the stochastic difference equation $\log x_{t+1} - (\log x_t + \varepsilon_t) = 0$. In continuous time $L(X) = 0$ will be a stochastic differential equation.

3.2 THE MARKOV DECISION MODEL

A form of Markov decision model provides a useful framework for introducing the symmetry concept. The usual Markov decision model splits variables into action variables and state variables. A slightly finer division proves helpful here. There is an endogenous state variable m that will usually be income or wealth. There are two types of action variables. The consumption-type variables c appear in the objective functional. Other actions that do not enter the objective, such as asset demands, are denoted by a. The capital stocks of the second example would fall into this category. The symmetry may act quite differently on these two types of actions, just as it could act differently on feasible sets and indifference curves in Figure 3. Finally, p includes all exogenous parameters. Prices played this role in the first example. These can be fixed, variable, or even stochastic. A parameter vector is explicitly included since more general symmetries may also act on parameters of the model.

Let $z = (m, c, a, p)$ denote a Markov process over the probability space Ω with index set $\mathfrak{I}$ that takes values in $\mathfrak{M} \times \mathcal{C} \times \mathcal{A} \times \mathcal{P}$.[5] The Markov process z is assumed to have evolution equations $L(z) = 0$. Various restriction on the values that state variables can take are summarized by the sets $\mathfrak{M}, \mathcal{C}, \mathcal{A}$ and $\mathcal{P}$. The economics of the problem will impose some structure on these sets. It may be important in the solution. If consumption must be non-negative, set $\mathcal{C} = \{c \in \mathbb{R}^n : c \geq 0\}$. When $p = p_0$ is fixed, take $\mathcal{P} = \{p_0\}$.

The objective functional defined over consumption paths will be denoted by V. Various types of objectives, such as Koopmans' (1960) recursive utility functions, are permitted. Typically, the objective functional V will be additively separable with immediate reward (*felicity*) $u(c, s)$. In that case, $V(c, t) = \int_t^T u(c, s) \, ds$ for continuous processes, and $V(c, t) = \Sigma_{s=t}^T u(c, s)$ for discrete processes.

Let E_t denote the conditional expectation given $m(t)=m_t$.[6] The basic problem can be written:

$$J(m_t, t \mid p) = \sup \{E_t V(c, t) : L(z) = 0\}.$$

I will refer to J as the *value function* or *indirect utility function*. For simplicity, I write $J(m_t, t)$ if p is fixed, and $J(m)$ if t is also fixed.

3.3 SYMMETRIES

To understand the geometric structure of the problem, focus on the budget set (feasible set). The *budget set* given initial state m_t and parameter p is $\mathbb{B}(m_t, t \mid p) = \{c \in \mathcal{C} : L(z) = 0 \text{ for some } z \text{ with } m(t) = m_t\}$. The important question is how the budget set's geometry depends on the initial state m_t. Transformations that leave the evolution equations invariant will be used to investigate this. These transformations map the budget set to another budget set. Let $T = T_1 \times T_2 \times T_3 \times T_4$ be an invertible (bijective) transformation on $\mathfrak{M} \times \mathcal{C} \times \mathfrak{A} \times \mathcal{P}$. It is a *symmetry* of the feasible set, provided that $L(Tz) = 0$ if and only if $L(z) = 0$.[7] Although the first problems I solve will use linear, time-independent symmetries, symmetries need not be linear. Sometimes the symmetries will be non-linear, sometimes they will be time-dependent, and sometimes they will even be stochastic. Later, I will even use symmetries that involve the exogenous parameters.

The following lemma shows the effect of a production symmetry on the budget set. The mapping T_2 maps any budget set to another budget set with possibly different initial data and parameters. This is the key fact used in the Symmetry Theorem.

Lemma. *If T is a symmetry then $\underline{\mathbb{B}}\,(T_1\,m_t\,/\,T_4\,p) = T_2\underline{\mathbb{B}}(m_t/p).$*

Proof. Let $c \in \underline{\mathbb{B}}\,(m_t\mid p)$. Take $z = (m, c, a, p)$ with $L(z) = 0$. Since T is a symmetry, $L(Tz) = 0$. Hence $T_2\,c \in \underline{\mathbb{B}}\,(T_1\,m_t\mid T_4 p)$, so $T_2\underline{\mathbb{B}}\,(m_t\mid p) \subset \underline{\mathbb{B}}(T_1 m_t\mid T_4 p)$.

As the T_i are invertible, we can apply the above result to $T'_i = T_i^{-1}$, $m'_t = T_1\,m_t$ and $p' = T_4 p$ to get $T'_2\underline{\mathbb{B}}(m'_t\mid p') \subset \underline{\mathbb{B}}(T'_1\,m'_t\mid T'_4 p')$. But $T'_1\,m'_t = m_t$ and $T'_4\,p' = p$, so $\underline{\mathbb{B}}(T_1 m_t\mid T_4 p) \subset T_2\underline{\mathbb{B}}(m_t\mid p)$. Hence $\underline{\mathbb{B}}\,(T_1 m_t\mid T_4 p) = T_2\underline{\mathbb{B}}(m_t\mid p).$

Q.E.D.

As in the finite-dimensional examples, the most helpful symmetries had to be symmetries for both preferences and the budget set. Since preferences are defined over Markov processes, some care must be used when looking for preference symmetries. The condition $E_t V(T_2 c, t) = f_t\,[E_t V(c, t)]$ for an increasing f_t insures that T will leave the indifference map invariant at each time t. Such a T is a preference symmetry. Generally, the expectation cannot be ignored here unless f_t is affine. When the budget symmetry is also a symmetry for preferences, this immediately yields information about the value function. Although it does not require that optima exist, the essence of the proof is in showing that optima are in fact mapped into optima.

Symmetry Theorem. *If $E_t V\,(T_2 c, t) = f_t\,[E_t V\,(c, t)]$ for some family of increasing functions f_t, then $J\,(T_1 m_t, t\mid T_4 p) = f_t\,[\,J\,(m_t, t\mid p)].$*

Proof. Since

$$J(T_1 m_t, t\mid T_4 p) = \sup\,\{E_t V(c, t) : c \in \underline{\mathbb{B}}\,(T_1 m_t\mid T_4 p)\}$$

we apply the lemma to get

$$J(T_1 m_t, t \mid T_4 p) = \sup \{E_t V(c, t) : c \in T_2 \mathbb{B} (m_t \mid p)\}$$

$$= \sup \{E_t V(T_2 c, t) : c \in \mathbb{B} (m_t \mid p)\}$$

$$= \sup \{f_t [E_t V(c, t)] : c \in \mathbb{B} (m_t, t \mid p)\}$$

Now since f_t is increasing, we can pull it through the supremum to get

$$J(T_1 m_t, t \mid T_4 p) = f_t [\sup \{E_t V(c, t)] : c \in \mathbb{B} (m_t, t \mid p)\}].$$

So

$$J(T_1 m_t, t \mid T_4 p = f_t [J (m_t, t \mid p)].$$

Q.E.D.

Note that the family of functions f_t can be time-dependent. In fact, it can even be stochastic.

4. APPLICATIONS TO PORTFOLIO PROBLEMS

Many portfolio problems are really no different from the examples given earlier. Consider a simple portfolio problem. It has two important characteristics—a linear homogeneous budget constraint and a homogeneous valuation functional defined over consumption paths. The same symmetry of uniform expansion still applies as in ordinary demand theory. The combination of a linear budget constraint and homogeneous valuation functional again gives rise to a homogeneous indirect utility function.

4.1 Specializations of the Basic Theorem

When m is a scalar variable, and L is linear homogeneous in (m, c), the Symmetry Theorem determines the form of the solution for additively separable V when $u(c, s)$ is either homogeneous or logarithmic in c.

Corollary 1. *Let m and c be real-valued and suppose V is an additively separable objective. Suppose $L(\lambda m, \lambda c, a, p) = \lambda L(m, c, a, p)$, $u(\lambda c, s) = \lambda^\alpha u(c, s)$ for $\lambda > 0$, and $\mathfrak{M}$ and $\mathcal{C}$ are cones ($\lambda \mathfrak{M} = \mathfrak{M}$, $\lambda \mathcal{C} = \mathcal{C}$). If $J(m_t, t \mid p)$ exists, it has the form $J(m_t, t \mid p) = A(t) \, m_t^a$ for some function A(t).*

Proof. Let $T_1 m = \lambda m$, $T_2 c = \lambda c$, $T_3 a = a$ and $T_4 p = p$ for $\lambda > 0$. Now $\lambda L(z) = L(Tz)$. Since these are simultaneously zero, T is a symmetry. Now, $V(T_2 c, t) = V(\lambda c, t) = \lambda^\alpha V(c, t)$ since $u(\lambda c, s) = \lambda^\alpha u(c, s)$. Set $f_t(V) = \lambda^\alpha V$. By the Symmetry Theorem,

$$J(\lambda m_t, t \mid p) = f[\, J\,(m_t, t \mid p)] = \lambda^\alpha J(m_t, t \mid p)$$

Now take $\lambda = 1/m_t$ and $A(t) = J(1, t \mid p)$ to get $J(m_t, t \mid p) = A(t) \, m_t^a$.
Q.E.D.

Corollary 2. *Under the conditions of Corollary 1 with $u(c, s) = f(s) \log c$, $J(m_t, t \mid p) = A(t) + F(t) \log m_t$, where $F(t) = \int_t^T f(s) \, ds$ in continuous time and $F(t) = \sum_{s=t}^T f(s)$ in discrete time.*

Proof. In continuous time, $V(\lambda c, t) = \int_t^T f(s) \log \lambda c \, ds = V(c, t) + F(t) \log \lambda$. Setting $f_t(V) = V + F(t) \log \lambda$, the Symmetry Theorem yields $J(\lambda m_t, t \mid p) = F(t) \log \lambda + J(m_t, t \mid p)$. Now set $\lambda = 1/m_t$ and $A(t) = J(1, t \mid p)$. This shows $J(m_t, t \mid p) = A(t) + F(t) \log m_t$. Discrete time is similar.
Q.E.D.

The key fact behind the corollaries is that the symmetry T is related to the objective in a simple way: $V(T_2 c, t) = V(\lambda c, t) = \lambda^\alpha V(c, t)$ or $V(T_2 c, t) = \log \lambda + V(c, t)$. Maxima are transformed into maxima. The optimal controls (c^*, a^*) for the problem with initial conditions $m(t) = m_t$, are transformed into the optimal controls $(T_2 c^*, T_3 a^*)$ for the problem with initial conditions $m(t) = T_1 m_t$. When the optimal controls exist, this fact can be quite useful. Hahn (1970) used it to determine optimal saving and consumption.

Symmetries can also help illuminate existing results. Danthine and Donaldson (1981) have shown that consumption is not a linear function of output in stochastic control problems with Cobb-Douglas production and homogeneous logarithmic felicity, although it is linear for logarithmic felicity. A close look at the appropriate symmetry explains why. In fact, this result even appeared in Section Two in a deterministic setting with two goods. Neither an infinite-horizon nor uncertainty are necessary for this result.

4.2 Merton's Portfolio Choice Model

An easy application is to a version of Merton's (1969, 1971) portfolio problem. In Merton's model, there are n assets with prices $P_1, P_2, ..., P_n$ described by Itô processes. These are most conveniently written in terms of the return to investment $dP_i/P_i = a_i(P, t) dt + \sigma_i(P, t) dz_i$. The z_i's are independent Brownian motions. Merton derives the budget constraint by considering a discrete-time version of the model. Assume that trading is done in periods of length h. The wealth level $m(t)$ and prices $P_i(t)$ are known at the beginning of period t, and are constant over the period. Denote the number of shares of asset i purchased in the interval $[t, t + h)$ by $N_i(t)$, and let $c(t)$ be the rate of consumption in the interval $[t, t + h)$. All wealth is held in the form of assets purchased in the previous period, so the initial wealth obeys

$$m(t) = \sum_{i=1}^{n} N_i(t - h) \, P_i(t).$$

Out of this, the consumer chooses a consumption rate and a new portfolio. Thus

$$c(t)h + \sum_{i=1}^{n} N_i(t) \, P_i(t) = \sum_{i=1}^{n} N_i(t - h) \, P_i(t) \qquad (1)$$

Prices change randomly at the end of the period, and the process iterates.

Applying (1) at time $t + h$ shows

$$c(t + h)h = \sum_{i=1}^{n} [N_i(t + h) - N_i(t)] \, P_i(t + h)$$

$$= \sum_{i=1}^{n} [N_i(t + h) - N_i(t)][P_i(t + h) - P_i(t)]$$

$$+ \sum_{i=1}^{n} [N_i(t + h) - N_i(t)] \, P_i(t)$$

Letting $h \to 0$ shows

$$-c(t) \, dt = \sum_{i=1}^{n} dN_i(t) \, dP_i(t) + \sum_{i=1}^{n} dN_i(t) \, P_i(t).$$

Similarly,

$$m(t) = \sum_{i=1}^{n} N_i(t) \, P_i(t)$$

An application of Itô's Lemma shows

$$dm = \sum_{i=1}^{n} N_i \, dP_i + \sum_{i=1}^{n} P_i \, dN_i + \sum_{i=1}^{n} dN_i \, dP_i$$

Thus

$$dm = -c(t) \, dt + \sum_{i=1}^{n} N_i(\mathrm{t}) \, dP_i$$

Set $a_i = N_i P_i / m$ and substitute for dP_i / P_i to get the budget constraint

$$dm = (\textstyle\sum_{i=1}^{n} \alpha_i a_i m - c)\, dt + \textstyle\sum_{i=1}^{n} a_i \sigma_i m\, dz_i \qquad (2)$$

Now consider a simple form of this model. In it, there are two assets—a safe asset that pays a certain return r, and a risky asset that pays a stochastic return with variance σ and expected return ρ. The total return from a unit of the risky asset in a time interval dt is $dP_r / P_r = [\rho\, dt + \sigma\, dz]$, while the return from a unit of the safe asset is $dP_s / P_s = r\, dt$. Total wealth is m and a is the share of wealth held in the risky asset. Income from the safe asset is $[(1 - a)mr\, dt]$ while income from the risky asset is $[am\rho\, dt + a\sigma\, dz]$. Since the consumption flow is c, the budget constraint can be written as the stochastic differential equation $L(m, c, a, p) = dm - a(\rho - r)m\, dt - (rm - c)\, dt - am\sigma\, dz = 0$ where $p = (r, \rho, \sigma)$. When short selling ($a < 0$ or $a > 1$) is permitted, the problem is

$$J(m_t, t) = \sup_{\{c,\, a\}} E_t \int_t^T c^\alpha e^{-\beta s} ds$$
$$\text{s.t.} \quad dm = L(m, c, a, p);$$
$$c \geq 0,\ m \geq 0;\ m(t) = m_t.$$

Here $\mathfrak{M} = \{m : m \geq 0\}$, $\mathcal{C} = \{c : c \geq 0\}$ and $\mathcal{A} = \{a\}$. If short selling were prohibited, take $\mathcal{A}' = \{a : 0 \leq a \leq 1\}$. In either case, Corollary 1 clearly applies. When the value function exists, it is given by $J(m_t, t) = A(t)\, m_t^\alpha$. Knowing the form of J in advance is quite useful. The Bellman equation can have extraneous solutions. Merton (1969) used a transversality condition to eliminate them. The symmetry approach automatically eliminates the extraneous solutions from consideration.

Many of the economically important results to not need the explicit solution for J, only the form determined by the corollaries. We can immediately see that the value function has absolute risk aversion $(1 - \alpha)/m$ and relative risk aversion $(1 - \alpha)$, regardless of what $A(t)$ is. This can be used in the first order conditions to show that consumption is linear in wealth and that the optimal asset share is $a = (\rho - r) / \sigma^2(1 - \alpha)$. If this is in α', this is also the solution to the problem without short selling. Merton (1969, 1971, 1973) has computer $A(t)$ in detail. When short selling is prohibited, and $(\rho - r) / \sigma^2(1 - \alpha)$ is not in α', all wealth must be held in only one of the assets ($a = 0$ or 1).

4.3. Applications of Merton's Model

The same symmetry works on closely related versions of this problem. Logarithmic felicity functions can be treated the same way by using Corollary 2 in place of Corollary 1. These results can also be extended to the case where there are many risky assets. One such model, involving two state equations, is due to Fischer (1975). He uses a variant of the Merton model where there are two risky assets with returns $dP_1 / P_1 = [\rho_1 dt + \sigma_1 dz_1]$ and $dP_2 / P_2 = [\rho_2 dt + \sigma_2 dz_2]$ and a safe asset with return $dP_3 / P_3 = \rho dt$. The price of the consumption good is given by a stochastic inflation process p with expected inflation rate π and variance σ_p. The consumer's wealth shares of the two risky assets are a_1 and a_2. This leaves $a_3 = (1 - a_1 - a_2)$ as the wealth share of the safe asset. Remembering that nominal consumption is pc, substitute in equation (2) to obtain the budget constraint $dm = [a_1 \rho_1 + a_2 \rho_2 + (1 - a_1 - a_2) r] m \, dt - pc \, dt + m(a_1 \sigma_1 dz_1 + a_2 \sigma_2 dz_2)$. Two distinct symmetries arise from the economic structure of the problem. By using both, I can extract more information than would otherwise be possible.

Fischer's problem is

$$J(m_0, P_0) = \sup_{\{c, a_i\}} E_0 \int_0^\infty u(c, s) \, ds$$

$$\text{s.t.} \quad dm = [a_1(\rho_1 - r) + a_2(\rho_2 - r)]m \, dt + (rm - pc) \, dt + m(a_1 \sigma_1 dz_1 + a_2 \sigma_2 dz_2)$$

$$dp = \pi p \, dt + p\sigma_p dz_p$$

$$m \geq 0, \, c \geq 0, \, m(0) = m_0, \, p(0) = p_0.$$

There is a second symmetry acting on the price level p in addition to the familiar symmetry $T_1 m = \lambda m$, $T_2 c = \lambda c$ and $T_3 a = a$. The second symmetry is $S_1(m, p) = (\lambda m, \lambda p)$, $S_2 c = c$ and $S_3 a = a$. This symmetry is particularly interesting since it leaves the budget set unchanged. It expresses the fact that the budget set depends only on real wealth, not nominal wealth. By the Symmetry Theorem, applied to $S = (S_1, S_2, S_3)$, $J(\lambda m_0, \lambda p_0) = J(m_0, p_0)$. Setting $\lambda = 1/p_0$ yields $J(m_0, p_0) = J(m_0/p_0, 1)$ $= I(m_0/p_0)$ for some function I. The value function J is also a function of real, not nominal, wealth.

We can use this to derive Fischer's important relation

$$J_{mp} p / J_{mm} m = -(J_m / J_{mm} m) - 1 \qquad \text{(equation 20 in Fischer)}.$$

Although Fischer uses this equation, he never actually shows it is true. Rather, he asserts that consumption depends on real wealth and uses a first order condition to get his equation (20). We can go the opposite way to see that Fischer's assertion is true. Use the first order condition $u_c(c, s) = p J_m$. As $p J_m = p(I'(m/p)/p) = I'(m/p)$, and since consumption satisfies $u_c(c, s) = I'(m/p)$, it is a function of real wealth.

In the special case where felicity is homogeneous (or logarithmic) we can use the standard symmetry T and the two corollaries to get, respectively, $J(m_0, p_0) = A(m_0/p_0)^\alpha$ and $J(m_0, p_0) = A + [\log(m_0/p_0)]$ $\times \int_0^\infty f(s) \, ds$.

The same type of symmetry can be applied to the asset pricing models with multiple consumption variables examined by Gertler and Grinols (1982). The Gertler-Grinols model has a somewhat more involved constraint that includes bonds B, money balances M, capital a_1 and real consumption c_1. The nominal return on bonds is certain and equal to i, while money has a nominal return of zero. Real output E evolves according to $dE = r a_1 \, dt + \sigma_1 a_1 \, dz_1$, while the price p of the consumption good follows $dp/p = \pi \, dt + \sigma \, dz$. Let $a_2 = B/p$ be the real bond holdings and $c_2 = M/p$ be real money balances. By Itô's Lemma,

$da_2/a_2 = (i + \sigma^2 - \pi) dt - \sigma dz$ and $dc_2 = (\sigma^2 - \pi) dt - \sigma dz$. In addition, each household receives a stochastic real transfer payment that is proportional to wealth and given by $dV = \theta m \, dt + m\sigma_s \, dz_s$. The total change in real wealth is then $dm = dE + da_2 + dc_2 + dV - c_1 \, dt$. Substituting and collecting terms yields the constraint $dm = [a_1 + (i + \sigma^2 - \pi)a_2 + (\sigma^2 - \pi)c_2 + \theta m - c_1] \, dt + \sigma_1 a_1 \, dz_1 - \sigma(a_2 + c_2) \, dz + m\sigma_s \, dz_s$. Their model also has money in the felicity function with felicity $u(c_1, c_2) = \alpha \log c_1 + (1 - \alpha) \log c_2$, so the problem is

$$
\begin{aligned}
J(m_t, t) = \sup_{\{c_i, a_i\}} \quad & E_t \int_t^\infty [\alpha \log c_1 + (1 - \alpha) \log c_2] \, e^{-\beta s} \, ds \\
\text{s.t.} \quad dm = & [a_1 + (i + \sigma^2 - \pi)a_2 + (\sigma^2 - \pi)c_2 + \theta m - c_1] \, dt \\
& + \sigma_1 a_1 \, dz_1 - \sigma(a_2 + c_2) \, dz + m\sigma_s \, dz_s \\
& c_i \geq 0, m \geq 0; \; m(t) = m_t
\end{aligned}
$$

An inspection of the budget constraint reveals a linear symmetry. It is $T_1 m = \lambda m$, $T_2(c_1, c_2) = (\lambda c_1, \lambda c_2)$ and $T_3(a_1, a_2) = (\lambda a_1, \lambda a_2)$. As this is also a symmetry for utility with $f_t(V) = V + (e^{-\beta s}/\beta) \log \lambda$, $J(\lambda m_t, t) = J(m_t, t) + (e^{-\beta s}/\beta) \log \lambda$, so $J(m_t, t) = A(t) + (e^{-\beta s}/\beta) \log m_t$. As in the case studied by Merton, this is sufficient to derive the all-important asset demands.

4.4 Symmetries With HARA Felicity

Another way to use symmetries is to transform a problem so that it is easier to solve. This works on Merton's (1971, 1973) portfolio problem for the HARA class function $u(c, s) = (c + \eta)^\alpha e^{-\beta s}/\alpha$ where $c + \eta \geq 0$, $\alpha < 1$ and $\alpha, \eta \neq 0$. The constraint on c, $c + \eta \geq 0$, is a bit different from the homogeneous Merton problem. The strategy is simple. Use a budget symmetry with $T_2 c = c + \eta$ to turn this into a homogeneous felicity function.

This sort of symmetry is a bit different from the previous examples because the constraint changes. The solution is to use a symmetry with $(T_1, T_2, T_3) : \mathfrak{M} \times \mathcal{C} \times \mathcal{C} \rightarrow \mathfrak{M}' \times \mathcal{C}' \times \mathcal{C}'$. The T_i must still be invertible and satisfy the budget constraint

$$dm = a(\rho - r)m \, dt + (rm - c) \, dt + am\sigma \, dz. \tag{3}$$

Let $m^* = T_1 m$ and $a^* = T_3 a$. For (T_1, T_2, T_3) to be a symmetry, the transformed variables must solve

$$dm^* = a^*(\rho - r)m^* \, dt + (rm^* - c - \eta) \, dt + a^*m^*\sigma \, dz.$$

Substituting for c from (3) gives us

$$dm^* = dm + (a^*m^* - am)[(\rho - r) \, dt + \sigma \, dz]$$
$$+ r(m^* - m)dt - \eta \, dt.$$

Choose a^* to eliminate the stochastic part of the equation. Letting $a^* = a(m/m^*)$ so that $a^*m^* = am$ yields

$$d(m^* - m) = r(m^* - m) \, dt - \eta \, dt. \tag{4}$$

This is a crucial point. Equation (4) has many solutions. Which one should we choose? Since $\mathfrak{M}' = \{m^* : m^* \geq 0\}$, our choice determines the constraint set $\mathfrak{M}$ for the inhomogeneous problem. One reasonable possibility is to require that terminal wealth $m(T)$ be non-negative. Under this condition, any borrowing must be repaid by time T. The boundary condition $m^*(T) = m(T)$ will insure this. This leads to Merton's solution. A different choice here would mean a different constraint set $\mathfrak{M}$.

With $m^*(T) = m(T)$ as boundary condition, the solution is

$$T_1 m = m^* = m + (\eta/r) \, [1 - e^{-r(T-t)}]$$

This gives $M = \{m: m + (\eta/r) [1 - e^{-r(T-t)}] \geq 0\}$. Of course, the value function is

$$J(m_t, t) = A(t) \left[m_t + (\eta/r) [1 - e^{-r(T-t)}] \right]^\alpha.$$

Another choice of constraint would give a different solution. The economic structure of the problem enters in a decisive way through this constraint.

In fact, this solution is quite different from the solution to the constant relative risk aversion case. Since log m^* is a Brownian motion, there is a positive probability of $\log m^* < \log \{(\eta/r) [1 - e^{-r(T-t)}]\}$ for $t < T$ when $\eta > 0$. It follows that m has a positive probability of being negative for $t < T$, even though terminal wealth $m(T) \geq 0$. Further, as $c^* = \theta m^*$ for some constant θ, there is also a positive probability of negative consumption. Further interpretation is required to make sense of this case. Merton works in a partial equilibrium setting, and c need only represent consumption out of wealth. Thus each agent may receive η units of the consumption good in each period, independent of wealth yielding total consumption $c + \eta$. An alternative is to require wealth and consumption to be non-negative. This path is followed by Sethi and Taksar (1989).

This symmetry has many interesting features. Most importantly, the symmetry no longer applies to a single problem, but transforms one problem into another. It also differs from the previous symmetries in other respects. It is affine, not linear, and the transformation is time-dependent. This last feature appears again with non-linear production functions.

5. SYMMETRIES AND NON-LINEAR PRODUCTION

Another application of symmetries is to models involving Cobb-Douglas production and logarithmic felicity. In this section, I use symmetries on a stationary Ramsey problem examined by Brock and

Mirman (1972) and Mirman and Zilcha (1975) and a non-stationary version of the same problem inspired by Mitra. The same method will also apply to more complex optimal growth models of this type, as those used in Radner (1966) and Long and Plosser (1983).

5.1 An Optimal Growth Model

The simple stochastic optimal growth model of Brock and Mirman (1972) contains a single all-purpose good which may be either consumed, or used as input in a production process. Let c_t, k_t and y_t respectively denote consumption, capital and income at time t. Uppercase letters will denote the corresponding infinite sequence, thus $C = \{c_t\}_{t=1}^{\infty}$. The production technology is described by a stochastic production function. For each value of a random variable ρ, the production function $f(k, \rho)$ is increasing and concave in k with $f(0, \rho) = 0$. The stochastic process influencing production $\{\rho_t\}$, is a sequence of independent random variables, and hence Markov. Thus income at time t is $y_t = f(k_{t-1}, \rho_{t-1})$, and consumption and the capital stock in period t are subject to the budget constraint $k_t + c_t \leq y_t = f(k_{t-1}, \rho_{t-1})$. Initial income is y in period 1.

A von Neumann-Morgenstern utility function is derived from a felicity function u which is increasing and concave. Utility is the discounted sum of felicity, with discount factor δ, $0 < \delta < 1$. Expected utility at time 1 is $E_0 [\sum_{t=1}^{\infty} \delta^{t-1} u(c_t)]$ where E_t denotes the expectation conditional on the value of ρ_t. Since ρ_0 is known at time 0 while ρ_1 is still unknown, the expectation is taken conditional on ρ_0. The optimal growth problem is to maximize expected utility subject to the feasibility constraint.

The variables of this model are easily placed in the Markov decision framework of Section Three. The state variable is y, c is an action variable that enters the objective, and k is an action variable that does not enter the objective. Everything else is a parameter.

5.2 Time Stationary Felicity

A simple stochastic Ramsey problem, studied by Mirman and Zilcha (1975), is

$$J(y) = \sup_{c_t} E_0 \left[\sum_{t=1}^{\infty} \delta^{t-1} \log c_t \right]$$

$$\text{s.t.}\quad c_t + k_t = (k_{t-1})^{\rho}{}_{t-1};\quad t = 2, 3, \ldots \tag{5}$$

$$c_t \geq 0, k_t \geq 0;\ c_1 + k_1 = y.$$

where k_t is capital stock at time t, $\delta < 1$ is the discount factor and the ρ_t are independent with mean $\rho < 1$.

For a symmetry to be useful, it must preserve the preference ordering. Consider the symmetries with $T_2 c_t = g_t(c_t)$ for a collection of functions g_t. These yield $V(T_2 C) = \sum_{t=1}^{\infty} \delta^{t-1} u(g_t(c_t))$. This representation is unique up to an increasing affine transformation (Debreu, 1969). Since logarithmic preferences are homothetic, Rader's results apply to show that either $u(g_t(c_t)) = a_t + \theta b_t \log c_t$ or $u(g_t(c_t)) = a_t + \theta_t(c_t)^\gamma$. The second case clearly does not preserve preferences. Further, unless b_t is constant, the discount factor is altered. This does not preserve preferences either, so set $b_t = 1$ and let $\log \alpha_t = a_t$. Then T_2 must have the form $T_2 c_t = \alpha_t c_t^{\theta}$ if it is a symmetry for preferences.

Let $T_1 k_t = k_t^*$ and $T_2 c_t = c_t^*$ be the transformation associated with $y^* = \lambda y$. Equation (5) must be satisfied by both the transformed and untransformed variables, so when $c_t = 0$, $k_t = (k_{t-1})^{\rho}{}_{t-1}$ and $k_t^* = (k^*_{t-1})^{\rho}{}_{t-1}$. Setting $z_t = k_t^*/k_t$, we see that z_t satisfies $z_t = (z_{t-1})^{\rho}{}_{t-1}$ with $z_1 = \lambda$. This has solution

$$z_t = \lambda_t \quad \text{where} \quad \lambda_t = \lambda^{(\prod_{i=1}^{t-1} \rho_i)}.$$

Notice that although ρ_t is stochastic, the transformation can be defined as easily as if it were deterministic.

Substituting the expressions for k_t^* and c_t^* into (5), we see that this is indeed a symmetry provided $\alpha_t = \lambda_t$ and $\theta = 1$. Applying the discrete-time analog of Theorem 1 shows that

$$J(\lambda y) = E_0 \left[\sum_{t=1}^{\infty} \delta^{t-1} \left(\prod_{i=1}^{t-1} \rho_i \right) \log \lambda \right] + J(y).$$

Setting $\lambda = 1/y$ and using the fact that ρ_t are independent with mean ρ, we get

$$J(y) = J(1) + [1/(1 - \rho\delta)] \log y.$$

The symmetries also show that the optimal policy function is linear in production y.

The same symmetry can be used on the undiscounted version of this problem (Mirman and Zilcha, 1977) to get

$$J(y) = J(1) + (1 - \rho)^{-1} \log y.$$

In general $J(1) \neq 0$, although Mirman and Zilcha erroneously give $J(1) = 0$. Without discounting, $J(1)$ is determined by normalizing J so that the Golden Rule has value zero. When ρ_t is deterministic ($\rho_t = \rho$), the Golden Rule initial endowment is $g = \rho^{\rho/(1-\rho)}$ and $J(1) = -(1 - \rho)^{-1} \log g$. Therefore, $J(y) = (1 - \rho)^{-1} \log (y/g)$.

5.3 Non-Stationary Felicity

An interesting variant on this problem was studied by Mitra (1979). Using a specialization of McKenzie's (1974) general time-varying model, Mitra considered a fixed felicity function with a time-varying discount rate. A further generalization, which I will examine, is to allow the technology to vary with time as well. For notational convenience, I only consider the deterministic case. These results may be extended to admit stochastic production as well. To insure that symmetries apply, I will consider Cobb-Douglas production and logarithmic felicity as above. Let $\gamma = (\gamma_1, \gamma_2, ...)$, $\rho = (\rho_1, \rho_2, ...)$, and $\Delta = (\delta_1, \delta_2, ...)$ be given. The technology at time t is given by $f_t(k_t) = \gamma_t(k_t)^{\rho_t}$ with δ_t representing the discount factor at time t. Then $y_t = f_{t-1}(k_{t-1})$ is income at time t. The Ramsey problem $P(y, \Delta, \rho, \gamma)$ becomes:

$$J(y \mid \Delta, \rho, \gamma) = \sup \ \textstyle\sum_{t=1}^{\infty} \ \delta^{t-1} \ \log c_t$$
$$\text{s.t.} \quad c_t + k_t = \gamma_{t-1} (k_{t-1})^{\rho_{t-1}} \quad \text{for } t = 2, 3, ...$$
$$c_1 + k_1 = y; \quad c_t, k_t \geq 0$$

Two symmetries combine to find the value function. The first transforms this into a problem with $\gamma_t = 1$. The symmetry $S(c_t, k_t) = (\exp \sigma_t)(c_t, k_t)$ does this when σ_t the difference equation $\sigma_t = \log \gamma_{t-1} + \rho_{t-1} \sigma_{t-1}$ with $\sigma_1 = 0$. The symmetry used on the Mirman-Zilcha model then does the rest.

To illustrate this, consider the case $\delta_t = \delta^{t-1}$, $\rho_t = \rho$ and $\gamma_t = e^{\gamma t}$. For the first symmetry, $\sigma_t = \gamma(t-1) + \rho\sigma_{t-1}$ and $\sigma_1 = 0$. As is easily verified, the solution to this difference equation is $\sigma_t = \gamma(\rho^t - \rho t + t - 1)$ $(1 - \rho)^{-2}$. The value function then obeys $J(y \mid \Delta, \rho, \gamma) = J(y \mid \Delta, \rho, 0) + \gamma\delta / [(1 - \delta)^2 (1 - \rho\delta)]$.

This transforms the problem into a deterministic Mirman-Zilcha model. Its solution yields $J(y \mid \Delta, \rho, \gamma) = A + \gamma\delta / [(1 - \delta)^2 (1 - \rho\delta)] + (1 - \rho\delta)^{-1} \log y$ where A is constant. An interesting point about this expression is that it permits computation of the optimal policy function, and hence facilitates explicit computation of the optimal path via the Bellman equation $J(y \mid \Delta, \rho, \gamma) = \sup \{\log c + \delta J (e^{\gamma}(y - c)^{\rho} \mid \Delta, \rho, \gamma)\}$.

Plugging in the expression for the value function, and using the first order condition, shows that the optimal choice of c_1 is $(1 - \rho\delta)y$. Similarly, $c_t = (1 - \rho\delta)y_t$ where y_t is the income available at time t. The optimal consumption path is then $c_t = \bar{c} \exp \{\gamma(\rho^t - \rho t + t - 1)$

$(1 - \rho)^{-2} + \rho^{t-1} \log (y/g)\}$ where $\bar{c} = (1 - \rho\delta)g$ and $g = (\rho\delta)^{\rho/(1-\rho)}$ denote steady state consumption and income, respectively.

Although technology grows at a constant rate, the growth rate of consumption varies. Asymptotically, the optimal path grows at rate $\gamma/(1 - \rho)$. At finite times, its behavior depends on the sign of $[\log (y/g) + \gamma\rho/(1 - \rho)^2]$. Consumption grows at an increasing, constant or decreasing rate as $[\log (y/g) + \gamma\rho/(1 - \rho)^2]$ is negative, zero or positive. In the first two cases, consumption is monotonically increasing in time. The last case involves another interesting possibility. With a large initial capital stock, the optimal consumption can be U-shaped. For example, with $\gamma = \rho = 1/2$, $0 < \delta < 1$ and $y = g^2 = \delta^2/4$, $c^t = \bar{c} \exp \{t + 3(1/2)^{t-1}\}$. Consumption in the first three periods is $c^1 = \bar{c} e^4$, $c^2 = \bar{c} e^{3.5}$ and $c^3 = \bar{c} e^{3.75}$.

Examination of the more general case $0 < \rho_t = \rho < 1$ with δ_t and γ_t arbitrary is also interesting. By the same procedure,

$$k^*_t = (1 - \delta_t/\mu_t)(1 - \delta_{t-1}/\mu_{t-1})^\rho \dots (1 - \delta_1/\mu_1)^{\rho^{t-1}} y^{\rho^{t-1}}$$

where $\mu_j = \sum_{t=j}^{\infty} \delta_t \rho^{t-j}$. Note that μ_j must be finite for all j if the problem is to be well posed. Let k^* have initial stock k_0 and z^* have initial stock k_1. Since $0 < (1 - \delta_t/\mu_t) < 1$,

$$| k^*_t - z^*_t | \leq | k_0^{\rho^{t-1}} - k_1^{\rho^{t-1}} | \to 0.$$

The optimal paths from different capital stocks are asymptotic to each other. This is known as the twisted turnpike. Mitra established this for the generic stationary technology case (1979). With time-varying production, the twisted turnpike need not hold. In fact, it does not generally hold with exogenous technical progress. This may be verified by applying l'Hôpital's rule to the explicit expression for $x_t = \rho\delta c_t/(1 - \rho\delta)$. To get a turnpike result we must first compensate for the growth rate. L'Hôpital's rule shows $| x_t - z_t | e^{-\gamma t/(1-\rho)} \to 0$.

6. EQUILIBRIUM MODELS

The symmetries of Section Five can be applied to equilibrium models with Cobb-Douglas production and logarithmic felicity. In such a setting, the same symmetry can be applied to all households. Provided a steady state is known, other equilibria may be investigated. One relatively simple equilibrium model is Michener's (1982) version of Lucas' (1978) asset pricing model. In addition to the value function, we also need to find the equilibrium pricing function. This causes Michener some mild embarrassment. Although Lucas did show that bounded utility functions have a unique equilibrium pricing function, his uniqueness theorem does not apply here. Fortunately Michener's answer is unique, and the symmetries can show it. Furthermore, even if we allow non-stationary prices à la Brock (1982; Milliaris and Brock, 1982), equilibrium prices are still unique. Michener's problem must be broken down into two parts. The first is solved conditional on asset prices. The second determines the equilibrium prices. The first problem is

$$
V(y, z \mid p) = \sup_{\{c, a\}} \; E_0 \left[\sum_{t=1}^{\infty} \delta^{t-1} \log c_t \right]
$$

$$
\text{s.t.} \quad \log y_{t+1} = \rho \log y_t + \varepsilon_t
$$
$$
c_t + p_t a_t = y_t z_t + p_t z_t; \quad z_{t+1} = a_t
$$
$$
z_t \geq 0, \, c_t \geq 0; \quad z_1 = z, \, y_1 = y.
$$

Of course $0 \leq \rho < 1$, $0 \leq \delta < 1$ and $\varepsilon_t \sim N(0, \sigma^2)$.

In each time period the consumer chooses a_t, the amount of the asset to hold at the end of the period, and consumption c_t. The consumer's asset holding is carried over and becomes the initial asset holding z_{t+1} for the next period. The asset pays a return equal to income y_t. A stochastic Cobb-Douglas production function controls the growth of income. The consumer also gets income from the sale of the current asset holdings at price p_t. Income is spent on consumption and asset holdings for the next period. The asset-pricing problem is to find a sequence of asset prices that induce the consumer to continue to hold the initial endowment of the asset.

To put this into the Markov framework, note that y and z are state variables, a is an action that doesn't enter the objective, c is an action that does enter the objective and p is a set of parameters. This problem, like Fischer's, has two sets of symmetries. The first, based on the linear budget constraint $c_t + p_t a_t = y_t z_t + p_t z_t$, is $T_1(y_t, z_t) = (y_t, \lambda z_t)$, $T_2 c_t = \lambda c_t$, and $T_3 a_t = \lambda a_t$. By Corollary 2, $V(y, \lambda z \mid p) = (1 - \delta)^{-1} \log \lambda + V(y, z \mid p)$. Hence

$$V(y, z \mid p) = V(y, 1 \mid p) + (1 - \delta)^{-1} \log z.$$

Further, if a_t^* is optimal from z, λa_t^* is optimal from λz at the same prices. Hence, p_t is not only an equilibrium price sequence for z, so that $a_t^* = z$, but is also an equilibrium price sequence for λz. Equilibrium prices are unaffected by changes in initial wealth z.

The second symmetry is a bit different since it also involves the price sequence. This symmetry is based on the Cobb-Douglas technology $\log y_{t+1} = \rho \log y_t + \varepsilon_t$, and is a deterministic version of the Mirman-Zilcha symmetry. It is $S_1(y_t, z_t) = (\lambda_t y_t, z_t), S_2 c_t = \lambda_t c_t$, $S_3 a_t = a_t$, and $S_4 p_t = \lambda_t p_t$ where $\lambda_t = \lambda^{\rho^t}$. This symmetry maps equilibrium prices into equilibrium prices since a_t is unchanged by it.

Proposition. *For each initial y, there is a unique equilibrium price sequence. It is given by the pricing function $p(y_t) = \delta(1 - \delta)^{-1} y_t$.*

Proof. An application of the Principle of Optimality shows that $q_t = p_{t+1}$ are equilibrium prices for (y_2, z). Another application yields

$$V(y, z \mid p) = \sup \{\log [yz + p_1(z - a_1)] + \delta E_0 V(y_2, a_1 \mid q)\}.$$

Now T shows that q_t are equilibrium prices for (y_2, z') for any z'. This symmetry also tells us $V(y_2, a_1 \mid q) = V(y_2, 1 \mid q) + (1 - \delta)^{-1} \log a_1$. Armed with this information, we now apply the first order conditions at $a_1 = z$ and find $p_1 = \delta(1 - \delta)^{-1}y$.

Of course, a similar argument applies to q_t. Hence, $p_2 = \delta(1 - \delta)^{-1}y_2$ also. A simple induction shows that $p_t = \delta(1 - \delta)^{-1}y_t$. This is necessarily unique.

Q.E.D.

The point is that the only equilibrium prices are actually given by the stationary equilibrium pricing function $p(y) = \delta(1 - \delta)^{-1}y$. Now let $V(y, z) = V(y, z \mid p_t(y))$ be the equilibrium value function. Using the Symmetry Theorem on S shows $V(\lambda y, z) = (1 - \rho\delta)^{-1} \log \lambda + V(y, z)$. Combining this with the results using T gives $V(y, z) = A + (1 - \delta)^{-1} \log z + (1 - \rho\delta)^{-1} \log y$ for some constant A.

As is usual with the symmetry technique, the actual nature of the stochastic term was irrelevant. We will get the same results whenever this problem is well-posed.

7. CONCLUSION

This paper has shown how symmetry arguments can be used to solve various kinds of maximization problems. In addition to the examples presented here, symmetries are able to solve various other problems. They can also be used on the exponential utility functions used by Holmstrom and Milgrom (1987) and Chang (1986). (These have an additive symmetry.) The type of symmetry used on the Mirman-Zilcha example can be applied to more general problems of the same type, such as Radner (1966) or Long and Plosser (1983). Prescott and Mehra's (1980; Mehra, 1984) recursive competitive equilibrium is another model where symmetries can be helpful.

Other types of equilibrium models may be examined. Becker's

(1980) Ramsey equilibrium, where agents may not borrow against future wage income is such an example (Boyd, 1986).[8] One interesting application of this is found by adding capital taxation to the model. With appropriate preferences and technology, symmetries can be used to calculate the transition path between steady states when tax rates are changed (Boyd, 1988). The symmetries give us an expression for each individual's utility, and can be used to analyze welfare.

Elementary techniques for finding symmetries were presented in the examples. In many cases, the question of existence of symmetries remains open. The Noether theorem should prove helpful for finding symmetries that apply to a given technology or preference order. It plays a key role in investigations of a related type of invariance in Sato (1981), Sato and Nôno (1983) and Logan (1977).

NOTES

1. Typical cases include Merton (1969, 1971), Samuelson (1969), Gertler and Grinols (1982), Malliaris and Brock (1982), and Michener (1982).
2. The major exceptions are Hahn (1970) and Mirrlees (1974) who use an embryonic form of the symmetry technique.
3. A general result for homogeneous felicity without uncertainty is due to Mino (1983). His proof involved manipulation of the Hamiltonian.
4. Weyl's (1952) book is a nice introduction to symmetry concepts and their use.
5. Each of the sets involved is a subset of some vector space.
6. Note that the conditional expectation is usually a random variable. For it to be a constant, m_t must be constant.
7. More generally, T is a symmetry between problems with evolution operators L and L' and constraints $\mathfrak{M} \times \mathcal{C} \times \mathfrak{A} \times \mathcal{P}$ and $\mathfrak{M}' \times \mathcal{C}' \times \mathfrak{A}' \times \mathcal{P}'$ provided $L(Tz) = 0$ if and only if $L'(z) = 0$ and T is an invertible transformation between $\mathfrak{M} \times \mathcal{C} \times \mathfrak{A} \times \mathcal{P}$ and $\mathfrak{M}' \times \mathcal{C}' \times \mathfrak{A}' \times \mathcal{P}'$. Such problems look the same in some abstract sense. Given appropriate preferences, the solution of one entails the solution of all of them. This expanded notion of symmetry will prove useful in Section Five.
8. The Ramsey equilibrium is further developed in Becker and Foias (1987) and Becker, Boyd, and Foias (1986).

REFERENCES

Arnold, Ludwig (1974), *Stochastic Differential Equations: Theory and Applications*, Wiley, New York.

Becker, Robert A. (1980), On the Long-Run Steady State in a Simple Dynamic Model of Equilibrium with Heterogeneous Households, *Quart. J. Econ. 95*, 375-382.

Becker, Robert A., John H. Boyd III and Ciprian Foias (1986), The Existence of Ramsey Equilibrium, Working Paper, Indiana University.

Becker, Robert A. and Ciprian Foias (1987), A Characterization of Ramsey Equilibrium, *J. Econ. Theory 41*, 173-184.

Boyd, John H., III (1986), *Preferences, Technology and Dynamic Equilibria*, PhD. Dissertation, Indiana University.

Boyd, John H., III (1988), Dynamic Tax Incidence with Heterogeneous Households, Working Paper, University of Rochester.

Brock, William A. (1982), Asset Prices in a Production Economy, in *The Economics of Information and Uncertainty* (J. J. McCall, ed.), University of Chicago Press, Chicago.

Brock, William A. and Leonard J. Mirman (1972), Optimal Economic Growth and Uncertainty: The Discounted Case, *J. Econ. Theory 4*, 479-513.

Chang, Fwu-Ranq (1986), A Theory of the Consumption Function: The Case of Non-capital Income Fluctuations, Working Paper, Indiana University.

Danthine, J.P. and John B. Donaldson (1981), Stochastic Properties of Fast vs. Slow Growing Economies, *Econometrica 49*, 1007-1033.

Debreu, Gerard (11960), Topological Methods in Cardinal Utility Theory, in *Mathematical Methods in the Social Sciences* (K. J. Arrow, S. Karlin and P. Suppes, eds.), Stanford University Press.

Fischer, Stanley (1975), The Demand for Index Bonds, *J. Pol. Econ. 83*, 509-534.

Gertler, Mark and Earl Grinols (1982), Monetary Randomness and Investment, *J. Mon. Econ. 10*, 239-258.

Hahn, Frank H. (1970), Savings and Uncertainty, *Rev. Econ. Stud. 37*, 21-24.

Holmstrom, Bengt and Paul Milgrom (1987), Aggregation and Linearity in the Provision of Intertemporal Incentives, *Econometrica 55*, 303-328.

Koopmans, Tjalling C. (1960), Stationary Ordinal Utility and Impatience, *Econometrica 28*, 287-309.

Logan, John D. (1977), *Invariant Variational Principles*, Academic, New York.

Long, John B., Jr. and Charles I. Plosser (1983), Real Business Cycles, *J. Pol. Econ. 91*, 39-69.

Lucas, Robert E., Jr. (1978), Asset Prices in an Exchange Economy, *Econometrica 46*, 1429-1455.

Malliaris, A. G. and William A. Brock (1982), *Stochastic Methods in Economics and Finance*, North-Holland, Amsterdam.

McKenzie, Lionel W. (1974), Turnpike Theorems with Technology and Welfare Function Variable, in *Mathematical Models in Economics* (J. Los and M. W. Los, eds.), North-Holland, New York.

Mehra, Rajnish (1984), Recursive Competitive Equilibrium: A Parametric Example, *Econ. Letters 16*, 273-278.

Merton, Robert C. (1969), Lifetime Portfolio Selection Under Uncertainty: The Continuous-Time Case, *Rev. Econ. Stat. 51*, 247-257.

Merton, Robert C. (1971), Optimum Consumption and Portfolio Rules in a Continuous-Time Model, *J. Econ. Theory 3*, 373-413.

Merton, Robert C. (1973), Erratum, *J. Econ. Theory 6*, 213-214.

Michener, Ronald W. (1982), Variance Bounds in a Simple Model of Asset Pricing, *J. Pol. Econ. 90*, 166-175.

Mino, Kazuo (1983), On the Homogeneity of Value Function of the Optimal Control Problem, *Econ. Letters 11*, 149-154.

Mirman, Leonard J. and Itzhak Zilcha (1975), On Optimal Growth Under Uncertainty, *J. Econ. Theory 11*, 329-339.

Mirman, Leonard J. and Itzhak Zilcha (1977), Characterizing Optimal Policies in a One-Sector Model of Economic Growth Under Uncertainty, *J. Econ. Theory 14*, 389-401.

Mirrlees, James A. (1974), Optimum Accumulation Under Uncertainty: The Case of Stationary Returns to Investment, in *Allocation Under Uncertainty* (Jacques H. Dreze, ed.), Wiley, New York.

Mitra, Tapan (1979), On Optimal Economic Growth with Variable Discount Rates: Existence and Stability Results, *International Econ. Rev. 20*, 133-145.

Prescott, Edward C. and Rajnish Mehra (1980), Recursive Competitive Equilibrium: The Case of Homogeneous Households, *Econometrica 48*, 1365-1379.

Rader, Trout (1981), Utility over Time, The Homothetic Case, *J. Econ. Theory 25*, 219-236.

Radner, Roy (1966), Optimal Growth in a Linear-Logarithmic Economy, *International Econ. Rev. 7*, 1-33.

Samuelson, Paul A. (1969), Lifetime Portfolio Selection by Stochastic Dynamic Programming, *Rev. Econ. Stat. 51*, 239-246.

Sato, Ryuzo (1981), *Theory of Technical Change and Economic Invariance: Application of Lie Groups*, Academic, New York.

Sato, Ryuzo and Takayuki Nôno (1983), *Invariance Prinicples and the Structure of Technology*, Springer, New York.

Sethi, Suresh P. and Michael Taksar (1989), A Note on Merton's "Optimum Consumption and Portfolio Rules in a Continuous-Time Model," *J. Econ. Theory, 46*, 395–401.

Weyl, Hermann (1952), *Symmetry*, Princeton University Press, Princeton.

CHAPTER 9

On Estimating Technical
Progress and Returns to Scale

Paul S. Calem

1. INTRODUCTION

Technological change and economies of scale are essential components of any theory of production. Production relationships may
exhibit increasing returns to scale over a range of outputs, and they
may be affected by technological processes that increase factor
productivity over time. Economists have undertaken empirical studies of technical progress and economies of scale for a variety of
industries and at various levels of aggregation. The approach generally taken in these studies is to estimate a specified production
relationship. In this paper, we examine two issues related to such
estimations: identification problems due to holotheticity, and estimation of the bias of technical change.

Sato (1980, 1981) has introduced the concept of a "holothetic
technology." Holotheticity of a technology implies that technical
progress and returns to scale will be indistinguishable, unless a priori
hypotheses are made concerning the structure of the returns to scale
function. In section 2 below, we further examine the implications of
holotheticity. We address the question: Under what hypotheses
concerning returns to scale will technical change and returns to scale
be identified if the technology is holothetic? A simple identification
requirement is presented. We also inquire as to whether misspecifi-

cation of the production structure can hide an identification problem. We find that it can, thereby leading to incorrect estimates of technical change and returns to scale.

Section 3 focuses on a related issue: how to estimate the bias of technical change along with returns to scale. Studies that have looked at both technical change and returns to scale generally have ignored the issue of technical change bias, i.e., the relative impact of technical change on the productivities of labor and capital. Common techniques for estimating returns to scale and technical change, such as the translog approximation, do not indicate the bias of technical change.[1] Section 3 presents formulas that can be used in conjunction with a production function estimation to compute the bias of technical change. Section 4 contains concluding remarks.

2. HOLOTHETICITY AND THE IDENTIFICATION OF TECHNICAL CHANGE

Sato's (1980, 1981) theoretical results concerning holotheticity of production under technical progress call into question the possibility of estimating technical progress along with returns to scale. Let us explore this issue somewhat further.

Following Sato, we represent the effect of technical progress on a production function $f(K, L)$ by a one-parameter family of transformations T_t (where t is a time parameter).[2] A production function is said to be *holothetic* under a family of technical progress transformations if the effect of technical progress is simply a relabeling of the isoquants of the production function. Formally $f(K, L)$ is *holothetic* under T_t if, for all t,

$$T_t \cdot f(K, L) = g(f(K, L), t) \tag{1}$$

Or, if the production structure is represented implicitly by a relation-ship $T_t \cdot f(K, L, Y) = 0$, we define $f(K, L, Y)$ to be *holothetic* under T_t if, for all t,

$$T_t \cdot f(K, L, Y) = g(f(K, L, Y / g(t))) \qquad (1')$$

The isoquant map of a holothetic production function is invariant under technical progress. Therefore, the effect of technical progress on a holothetic production function is indistinguishable from a scale effect. Applying the mathematical theory of Lie transformation groups, Sato shows how to determine which technologies will be holothetic under any given family of technical progress transformations.

In the absence of *a priori* hypotheses concerning the structure of returns to scale, holotheticity rules out the separate identification of technical progress and returns to scale. But usually it is plausible to restrict the form of a returns to scale function. Returns to scale may be hypothesized to yield U-shaped average costs, or its functional form may be specified. When does *a priori* specification of returns to scale enable identification of returns to scale and technical progress?

Suppose one is estimating a functional form that includes a speci-fication of returns to scale and that provides an adequate representa-tion of the underlying technology. Under this scenario, estimates of technical change and returns to scale will be obtainable if a simple identification condition is satisfied. This identification requirement is derived as follows:

Let $f(K, L)$ be the production function; let $[K(t), L(t) : -\infty \leq t \leq \infty]$ be the expansion path for capital and labor;

$$K(t) = K(0) + \kappa_1 t + \kappa_2 t^2 + \ldots; \quad L(t) = L(0) + \lambda_1 t = \lambda_2 t^2 + \ldots, \quad (2)$$

and let $F[f]$ be the returns to scale function. There will exist a family of transformations S_t such that

$$F[f(K(t), L(t))] = S_t \cdot f(K(0), L(0)), \text{ for all } t. \tag{3}$$

That is, the returns to scale associated with the given time path of capital and labor can be conceptualized as a family of transformations of a technology existing at time zero. To see that this is so, let $f(t) \equiv f(K(t), L(t))$ and let $t^*(\phi)$ satisfy $f(t^*(\phi)) = \phi$. For any t, and for any ϕ such that $t^*(\phi)$ is well defined, we specify

$$S_t \cdot \phi \equiv F[f(t + t^*(\phi))] \tag{4}$$

If there exists no t^* such that $\phi = f(t^*)$, then we set $S_t \cdot \phi = \phi$ for all t. Then, since $t^*(f(0)) = 0, S_t \cdot f(K(0), L(0)) = F[f(t)] = F[f(K(t), L(t))].$[3]

Thus, given observations $K(t)$, $L(t)$, and $Y(t)$ on capital, labor and output, we can write

$$Y(t) = \varepsilon T_t \cdot S_t \cdot f(K(0), L(0)), \tag{5}$$

where ε is a random error term. Now suppose that returns to scale $S_t \cdot f(K, L)$ is a function of the form $h(f, t, \kappa, \lambda, \alpha)$, where $\kappa = (\kappa_1, \kappa_2, ...)$ and $\lambda = (\lambda_1, \lambda_2, ...)$, and where α is a parameter vector to be estimated. Suppose also that the production function is holothetic so that $T_t \cdot f(K, L)$ is a function $g(f, t, \beta)$, where β is a parameter vector to be estimated. Then equation (5) becomes

$$Y(t) = \varepsilon \, g(h(f(0), t, \kappa, \lambda, \alpha), t, \beta). \tag{6}$$

We thus obtain the identification requirement:

The separate estimation of technical progress will be possible if the production structure is not holothetic. Otherwise, the separate estimation of technical progress and returns to scale will be possible if and only if α and β are not underidentified in equation (6).

As an example, let $f(K, L) = AK^\gamma L^{(1-\gamma)}$, $F[f] = f^\alpha$, and $T_t \cdot f(K, L)$ $= f(e^{\beta t}K, e^{\beta t}L)$. In addition, let the observed time paths of capital and labor be exponential:

$$K(t) = K(0)\, e^{\kappa t + X(t)}, \quad L(t) = L(0)\, e^{\lambda t + Z(t)}, \quad t = 0, 1, \ldots, \quad (7)$$

where $X(t)$ and $Z(t)$ are "residuals." Then equation (6) becomes

$$\ln Y(t) = [\ln A + \alpha\gamma \ln K(0) + \alpha(1 - \gamma) \ln L(0)$$
$$+ \alpha\gamma X(t) + \alpha(1 - \gamma) Z(t) + [(\kappa + \beta) \alpha\gamma \qquad (8)$$
$$+ (\lambda + \beta) \alpha(1 - \gamma)]t + \ln \varepsilon.$$

In equation (8), α and β are identified, so long as both $X(t)$ and $Z(t)$ are stochastic. The efficiency with which these parameters can be estimated will depend upon the degree of variability in $X(t)$ and $Z(t)$.

Thus far, we have shown that holotheticity of a technology does not necessarily rule out estimation of technical progress and returns to scale. Moreover, so long as the estimated functional form is not a misspecification of the true production structure, a simple identification requirement indicates whether efficient estimation of technical progress and returns to scale is feasible. But what if returns to scale or some other aspect of the underlying technology has been misspecified? In this case, it turns out that the estimated production function may satisfy the above identification requirement though the true production structure does not. Hence, in this case, false—though seemingly reliable—estimates of technical progress and returns to scale could be obtained.

To see how such a problem might arise, consider the following example. Suppose that the observed time path of output satisfies

$$Y(t) = Y(0) + \delta_1 t + \delta_2 t^2 + \eta: \qquad \delta_i > 0;\ i = 1, 2 \qquad (9)$$

where η is white noise; and suppose that the true production relationship is given by

$$\ln Y(t) + \alpha_1 Y(t) = \ln A + \alpha_2 \{ \gamma [\ln K(t)] + (1-\gamma) \ln L(t)$$
$$+ \beta_1 t + \beta_2 t^2 \} + \varepsilon \qquad (10)$$

In equation (10), $\alpha_i > 0$ are returns to scale parameters and $\beta_i > 0$ are technical progress parameters. Zellner and Revanker (1969) show that for a production function of the form (10), returns to scale at Y equals $\alpha_2 / [1 + \alpha_1 Y]$; that is, returns to scale falls from α_2 at $Y = 0$ to zero as $Y \to \infty$. However, a researcher employing a translog specification would overestimate returns to scale to be approximately α_2, and would underestimate technical progress. That is, the translog estimation would capture the decline in economies of scale as a slow-down in the rate of technical change.

The failure of the translog estimation in the preceding example reflects the failure of the underlying production structure to satisfy our identification requirement. The example points to the importance in productivity estimations of checking for such specification error, by considering whether the underlying production structure might fail to satisfy our identification requirement. The point of departure for such an analysis would be a comprehensive listing of holothetic technologies. Sato (1981) has provided such a listing (along with a method for deriving holothetic technologies). The next step would be to check whether the observed data is consistent with some holothetic technology that, for a "reasonable" specification of returns to scale, violates the above identification requirement. How one best could go about making such a determination is a topic calling for further research.

3. ESTIMATING THE BIAS OF TECHNICAL CHANGE

Commonly utilized procedures for estimating technical progress along with returns to scale do not yield estimates of the bias of technical change. Consider, for instance, a translog approximation to a production function $F(A(t)K, B(t)L)$. By estimating the translog function, one can obtain an estimate of returns to scale, $\partial \ln F / \partial \ln K + \partial \ln F / \partial \ln L$, and an estimate of technical change, $\partial \ln F / \partial t$. However the bias of technical change—for example, the individual rates of change $d \ln A / dt$ and $d \ln B / dt$—will not be obtainable directly from such an estimation procedure. The following proposition presents identities that can be used, in conjunction with certain estimation procedures, to obtain consistent estimates of the bias of technical change. These equations generalize Sato's (1970) equations for estimating the bias of technical progress when production is linear homogeneous.

Proposition 1.   Suppose that $Y(t) = F[f(A(t) K(t), B(t) L(t))]$, where f is a linear homogeneous production function with biased technical progress, $F[f]$ is a returns to scale function, and $K(t), L(t)$, and $Y(t)$ are time paths for capital, labor, and output. In addition, let $\sigma(t)$ denote the elasticity of substitution of the function f, evaluated at t, and let $r(t)$ and $w(t)$ denote $\partial F / \partial K$ and $\partial F / \partial L$, respectively, evaluated at t. Finally, let $h(t)$ denote dY / df evaluated at t, and define $\alpha(Y) \equiv d\ln Y / d\ln f$. Then

$$
\frac{d \ln A}{dt} = \left[\frac{1}{\sigma(t)-1} \right] \left[\sigma(t) \left(\frac{d \ln w}{dt} - \frac{d \ln h}{dt} \right) - \left(\frac{1}{\alpha(Y)} \right) \left(\frac{d \ln Y}{dt} \right) + \frac{d \ln L}{dt} \right] \tag{11}
$$

$$
\frac{d \ln B}{dt} = \left[\frac{1}{\sigma(t)-1} \right] \left[\sigma(t) \left(\frac{d \ln r}{dt} - \frac{d \ln h}{dt} \right) - \left(\frac{1}{\alpha(Y)} \right) \left(\frac{d \ln Y}{dt} \right) + \frac{d \ln K}{dt} \right] \tag{12}
$$

The proof is in Appendix A.[4]

Equations (11) and (12) can be simplified in the case that the production function is homogeneous. In this regard, note that $h(t) = dY / df = F'(f) = \alpha(Y)Y / f$ and hence

$$\frac{d \ln h}{dt} = \frac{d \ln \alpha(Y)}{dt} + \frac{d \ln Y}{dt} - \frac{d \ln f}{dt} \qquad (13)$$

But $d \ln f / dt = [1 / \alpha(Y)] d \ln Y / dt$. Therefore, equation (13) implies

$$\frac{d \ln h}{dt} = \left(\frac{1}{\alpha(Y)} \right) \left(\frac{d\alpha(Y)}{dt} \right) + \left(\frac{\alpha(Y)-1}{\alpha(Y)} \right) \left(\frac{d \ln Y}{dt} \right) \qquad (14)$$

In the case that the production function is homogeneous of degree α, $\alpha(Y)$ is constant and equation (14) becomes $d\ln h / dt = [(\alpha(Y)-1)/ [\alpha(Y)] (d\ln Y / dt)$. In this case, equations (11) and 12) reduce to

$$\frac{d \ln A}{dt} = \left[\frac{1}{\sigma(t)-1} \right] \left[\sigma(t) \left\{ \frac{d \ln w}{dt} - \left(\frac{\alpha-1}{\alpha} \right) \left(\frac{d \ln Y}{dt} \right) \right\} \right.$$
$$\left. - \left(\frac{1}{\alpha} \right) \left(\frac{d \ln Y}{dt} \right) + \frac{d \ln L}{dt} \right] \qquad (11')$$

$$\frac{d \ln B}{dt} = \left[\frac{1}{\sigma(t)-1} \right] \left[\sigma(t) \left\{ \frac{d \ln r}{dt} - \left(\frac{\alpha-1}{\alpha} \right) \left(\frac{d \ln Y}{dt} \right) \right\} \right.$$
$$\left. - \left(\frac{1}{\alpha} \right) \left(\frac{d \ln Y}{dt} \right) + \frac{d \ln K}{dt} \right] \qquad (12')$$

There are alternative ways of applying proposition 1 to obtain estimates of the bias of technical change.[5] One approach involves adapting Sato's (1970) procedure. According to this approach, $d \ln A / dt$ and $d \ln B / dt$ are assumed to be constant. In addition, a data series for $w(t)$ and $r(t)$ is constructed by assuming that these variables equal the wage rate and the rate of return on capital, respectively. Finally, the production function is assumed to be homothetic CES or CEDD, so that the elasticity of substitution is a function of one unknown parameter. Equations (11) and (12) then serve as parameter restrictions that make estimation of the production structure feasible.

An alternative procedure is to obtain estimates of $r(t)$, $w(t)$, $\sigma(t)$, $\alpha(Y)$, and $d\ln h/dt$. These estimates can be obtained from the estimated parameters of a translog or related function:

$$F(Y) = a_1 \ln K + b_1 \ln L + a_2(\ln K)^2 + b_2(\ln L)^2 + c \ln K \ln L + d_1 t + d_2 t^2$$

$+ e_1 t \ln K + e_2 t \ln L$, by applying the following proposition.

Proposition 2.   *Let* $Y(t) = F[f(K(t), L(t))]$, *where f is linear homogeneous, and let* $\pi = (\partial \ln Y / \partial \ln L) / (\partial \ln Y / \partial \ln K)$. *Then*

$$r(t) = \left(\frac{\partial \ln Y}{\partial \ln K} \right) \left(\frac{Y(t)}{K(t)} \right); \quad w(t) = \left(\frac{\partial \ln Y}{\partial \ln L} \right) \left(\frac{Y(t)}{L(t)} \right) \quad (15)$$

$$\alpha(Y) = \left(\frac{\partial \ln Y}{\partial \ln K} \right) + \left(\frac{\partial \ln Y}{\partial \ln L} \right) \quad (16)$$

$$\frac{d \ln h}{dt} = \left(\frac{1}{\alpha(Y)} \right) \left(\frac{d\alpha(Y)}{dt} \right) + \left(\frac{\alpha(Y)-1}{\alpha(Y)} \right) \left(\frac{d \ln Y}{dt} \right) \quad (17)$$

$$\sigma(t) = \frac{\pi}{\pi + \partial\pi / \partial \ln K} \quad (18)$$

Proof.   *Equation (15) follows immediately from the definitions* $r(t) \equiv \partial Y / \partial K$ *and* $w(t) \equiv \partial Y / \partial L$. *Equation (16) follows from Euler's theorem and the definition of* $\alpha(Y)$. *Equation (17) is the same as (14), derived above, while equation (18) is derived in Appendix B.*

Formulas (15) through (18) allow consistent estimates of $w(t)$, $r(t)$, $\alpha(t)$, $d\ln h/dt$, and $\sigma(t)$ to be computed from the estimated parameters of a flexible functional form. These estimates, in turn, can be inserted in equations (11) and (12) to obtain consistent estimates of the bias of technical change in each period.[6] Of course, this procedure is applicable only under the condition that the underlying production structure is homothetic. In the case of a translog functional form, the usual homogeneity restrictions would apply.

4. CONCLUSION

A critical factor in estimations of technical progress and returns to scale is whether the underlying production structure is holothetic. The isoquant map of a holothetic production function is invariant under technical progress. Hence, in the absence of *a priori* hypotheses concerning the structure of returns to scale, holotheticity rules out identification of technical progress and returns to scale. However, given an *a priori* specification of the returns to scale function, holotheticity is not necessarily an obstacle. In this case, a simple identification condition indicates whether estimates of technical progress and returns to scale are obtainable, with one proviso. The estimated production structure must be adequate representation of the underlying technology.

Lack of identification may be hidden by a misspecified production structure. That is, the estimated production structure may yield identification of technical progress and returns to scale even though the true production structure does not. False, though seemingly accurate, estimates would then be obtained. Further research is needed into ways of checking for such specification error.

A related issue is how to estimate the bias of technical change along with returns to scale. Common estimation procedures, such as the translog approximation, yield estimates of returns to scale and the overall rate of technical change. However, they do not indicate the bias of technical change. Information concerning technical change bias is important for understanding patterns of productivity growth. We have presented formulas that can be used, in conjunction with a production function estimation, to compute the bias of technical change along with returns to scale.

APPENDIX A

We prove proposition 2. By assumption,

$$Y(t) = F[f(A(t) K(t), B(t) L(t))] \qquad \text{(A-1)}$$

where f is linear homogeneous. Hence

$$\frac{d \ln Y}{dt} = \left(\frac{\partial \ln Y}{\partial \ln K} \right) \left(\frac{\partial \ln K}{\partial t} + \frac{\partial \ln A}{\partial t} \right) +$$

$$\qquad \text{(A-2)}$$

$$\left(\frac{\partial \ln Y}{\partial \ln L} \right) \left(\frac{\partial \ln L}{\partial t} + \frac{\partial \ln B}{\partial t} \right)$$

Since f is linear homogeneous, we can write
$f(A(t) K(t), B(t) L(t)) \equiv B(t) L(t) \bar{f}(x)$, where $x \equiv A(t) K(t) / B(t) L(t)$.
Then $r(t) \equiv \partial Y/ \partial K = A(t)(dY / df) (d\bar{f}(x) / dx) \equiv (d\bar{f}(x) / dx) A(t)h$.
Hence:

$$\frac{d \ln r}{dt} = \frac{d \ln A}{dt} + \frac{d \ln h}{dt} +$$

$$\qquad \text{(A-3)}$$

$$\left[\frac{(d^2\bar{f}(x) / dx^2) x}{d\bar{f}(x) / dx} \right] \left[\frac{d \ln A}{dt} - \frac{d \ln B}{dt} + \frac{d \ln K}{dt} - \frac{d \ln L}{dt} \right]$$

Now, since $F[f]$ is homothetic, its elasticity of substitution σ satisfies

$$\sigma = \frac{d\bar{f}(x)}{dx} \left[\frac{x (d\bar{f}(x) / dx) - \bar{f}(x)}{(d^2\bar{f}(x) / dx^2) x\bar{f}(x)} \right] \qquad \text{(A-4)}$$

See, for instance, Clemhout (1968). Together, equations A-3 and A-4 imply

$$\frac{d \ln r}{dt} = \frac{d \ln A}{dt} + \frac{d \ln h}{dt} + \tag{A-5}$$

$$\left(\frac{\gamma}{\sigma} \right) \left[\frac{d \ln A}{dt} - \frac{d \ln B}{dt} + \frac{d \ln K}{dt} - \frac{d \ln L}{dt} \right]$$

where $\gamma \equiv [\, x \left(\dfrac{d\bar{f}(x)}{dx} \right) - \bar{f}(x)] / \bar{f}(x) = \dfrac{\partial \ln f}{\partial \ln L} = \dfrac{\partial \ln Y / \partial \ln L}{\alpha(Y)}$

By equations A-2 and (13), equation A-5 reduces to

$$\frac{d \ln r}{dt} = \frac{d \ln A}{dt} + \frac{d \ln h}{dt} + \tag{A-6}$$

$$\left(\frac{1}{\sigma \alpha(Y)} \right) \left[\frac{d \ln Y}{dt} - \frac{\alpha(Y) \, d \ln K}{dt} - \frac{\alpha(Y) \, d \ln A}{dt} \right]$$

Rearranging terms yields

$$\frac{d \ln A}{dt} = \left(\frac{1}{\sigma - 1} \right) \left[\sigma \left\{ \frac{d \ln r}{dt} - \frac{d \ln h}{dt} \right\} - \frac{1}{\alpha(Y)} \frac{d \ln Y}{dt} + \frac{d \ln K}{dt} \right]$$

$$\tag{A-7}$$

Similarly we can derive

$$\frac{d \ln B}{dt} = \left(\frac{1}{\sigma - 1} \right) \left[\sigma \left\{ \frac{d \ln w}{dt} - \frac{d \ln h}{dt} \right\} - \frac{1}{\alpha(Y)} \frac{d \ln Y}{dt} + \frac{d \ln L}{dt} \right]$$

$$\tag{A-8}$$

Q.E.D.

APPENDIX B

We derive equation (18). The elasticity of subtitution $\sigma(t)$ is defined by

$$\sigma(t) \equiv \frac{K(t)f_K + L(t)f_L}{K(t)L(t)\left[-f_{LL}\dfrac{f_K}{f_L} + 2f_{KL} - f_{KK}\dfrac{f_L}{f_K}\right]} \tag{B-1}$$

Defining $\eta = \dfrac{f_L}{f_K}$, note that

$$\eta_K - \frac{\eta_L}{\eta} = \frac{1}{f_K}\left[f_{LK} - f_{KK}\frac{f_L}{f_L} - f_{LL}\frac{f_K}{f_K} + f_{LK}\right] \tag{B-2}$$

Hence,

$$\sigma(t) = [1 + \eta L(t)/K(t)] / \{L(t)[\partial\eta/\partial K - (\partial\eta/\partial L)/\eta]\} \tag{B-3}$$

Letting $\pi \equiv \eta L(t)/K(t) = (\partial \ln f/\partial \ln L)/(\partial \ln f/\partial \ln K)$, note that $\partial\pi/\partial K = [(\partial\eta/\partial K)L(t)/K(t)] - \eta L(t)/(K(t))^2$ and $\partial\pi/\partial L = [(\partial\eta/\partial L)L(t)/K(t)] + \eta/K(t)$. Hence, equation (B-3) reduces to

$$\sigma(t) = [1 + \pi]/[1 + \pi + \partial\pi/\partial\ln K - (\partial\pi/\partial\ln L)/\pi]. \tag{B-4}$$

But $\partial\pi/\partial\ln K = -\partial\pi/\partial\ln L$, since the production structure is homothetic. Thus, equation (B-4) becomes

$$\sigma(t) = [1 + \pi]/[1 + \pi + (1 + 1/\pi)\partial\pi/\partial\ln K], \tag{B-5}$$

which, after some manipulation, reduces to equation (18).

Q.E.D.

NOTES

1. As noted by Binswanger (1974), the bias of technical change will be evident from changes in factor shares if and only if the factor price ratio remains constant.
2. In addition, Sato assumes that the family of transformations T_t is a Lie transformation group; that is, the successive performance of two transformations belonging to the family is also a member of the family; there exists an identity transformation; and the inverse of any member of the family also belongs to the family.
3. It is easily verified that the family of transformations S_t thus defined will constitute a Lie Group.
4. The function $\alpha(Y)$, as defined in propostion 2, represents returns to scale at Y. See Zellner and Revankar (1969).
5. These methods place certain *a priori* hypotheses on the structure of production, and thereby circumvent the "impossibility theorem" of Diamond, McFadden, and Rodriquez (1978). That theorem concerns the nonidentifiability of the elasticity of substitution and bias of technical change in the absence of such *a priori* hypotheses.
6. A technique described in Kendall, Stuart, and Ord (1987) for computing the variance of functions of random variables, can be used to compute the variances and standard errors of $w(t)$, $r(t)$, etc., and in turn of $d \ln A(t) / dt$ and $d \ln B(t) / dt$.

REFERENCES

Binswanger, H. (1974). The measurement of technical change biases with many factors of production. *American Economic Review, 64*, 964–976.

Clemhout, S. (1968). The class of homothetic isoquant production functions. *Review of Economic Studies, 35*, 91–104.

Diamond, P., McFadden, D., and Rodriquez, M. (1978). Measurement of the elasticity of factor substitution and the bias of technical change. In M. Fuss and D. McFadden (eds.) *Production economics: A dual approach to theory and applications, (2)*. Amsterdam: North-Holland Publishing Company.

Kendall, M., Stuart, A. and Ord, J. (1987). *Kendall's advanced theory of statistics, (1)*. New York: Oxford University Press.

Sato, R. (1970). The estimation of biased technical progress and the production function. *International Economic Review, 11*, 179–208.

Sato, R. (1980). The impact of technical progress on the holotheticity of production functions. *Review of Economic Studies, 47*, 767–776.

Sato, R. (1981). *The theory of technical change and economic invariance: Application of Lie groups*. New York: Academic Press.

Zellner, A., and Revanker, N. (1969). Generalized production functions. *Review of Economic Studies, 36*, 241–250.

CHAPTER 10

On the Functional Forms of
Technical Change Functions[*]

Thomas M. Mitchell[**]

1. INTRODUCTION

In empirical studies of producer behavior and technical change, the role of technical change has largely been that of an unexplainable "residual." In the total factor productivity approach to technical change under an assumption of constant returns to scale, the rate of technical change is computed as the growth rate of real output less the share-weighted sum of the growth rates of the real factor inputs. In the dual cost function approach to producer behavior, the negative of the rate of technical change is computed as the growth rate of minimized cost less the cost-share-weighted sum of the growth rates of factor prices. (Under a maintained hypothesis of constant returns to scale, "minimized cost" can be replaced with the unit price of output.) Among all studies of producer behavior and technical change, however, there have been few attempts to explain or model the residually-computed rate of technical change. Some studies have "modeled" productivity change so that the "biases of technical change" could be

[*]An earlier version of this paper was presented at the annual meeting of the Society for Economic Dynamics and Control, Tempe, Arizona, March 9–11, 1988.

[**]The author has benefited from discussions with, and comments from, Tom Russell, Rolf Färe, Dan Primont, Akira Takayama, Rick Grabowski, Mary Norris, and Ryuzo Sato but is solely responsible for any remaining errors.

estimated for each factor, but this is not an attempt to discover the *nature* of technical change so much as it is an attempt to simply identify and interpret the *effects*. Indeed, as noted by Stevenson (1980, p. 162), "it appears to be a much easier task to categorize the direction of technological bias than to explain it."

The purpose of this paper is to characterize a broad class of technical change functions which belong to the family of "one-parameter groups of transformations." These technical change transformations can be embedded in standard econometric models of producer behavior with technical change and will permit a more rigorous analysis of technological bias than a simple determination of its direction. In addition, the introduction of technical change functions of the class studied here will allow an escape from the simple time trend econometric models which unfortunately, as Baltagi and Griffin (1988) observe, remain the norm. While econometric models are not necessary for the study of all issues concerning technical change, Baltagi and Griffin (1988, pp. 24-25) point out that an argument which supports the econometric approach is that parameter estimates of substitution and scale elasticities are "important in their own right, and a properly specified measure of technical change will reduce the bias in these measures."

Before proceeding we should ask: What is the value of modeling technical change functions as one-parameter groups of trans-formations? After all, technical change was studied before Sato's (1980, 1981) introduction of group theory to economics, and there is no reason to expect that group theory will be the final word in this area. It seems that there are at least three responses to the question. First, it is not a departure from existing work in the sense that radically new structure is being given to technical change functions. Sato (1980, p. 775, Note 6; 1981, p. 27) has noted that the technical change functions which have appeared in the literature actually have the group structure.

Second, as noted above, the class of technical change functions studied here will allow econometric models to escape the restrictive time trend form. With the existence of efficient nonlinear estimation procedures, this is a contribution of significant value.

Finally, in sharp contrast to the problem of modeling the technologies themselves, if we impose the group structure on technical change functions, characterization theorems, such as the theorem below, are possible. As Lewbel (1987b, p. 311) has observed in the area of consumer demand theory:

> Standard axioms of consumer preferences lead to demand systems having the properties of homogeneity, monotonicity, adding up, symmetry, and quasiconcavity. There is an unlimited number of model specifications satisfying these properties, resulting in a similarly endless stream of publications that propose and estimate new ones. Far fewer papers have been devoted to characterizing broad classes of demand systems. Instead of inventing new systems, characterization theorems define and delimit what forms of models can be devised.

This paper is devoted to characterizing a broad class of technical change transformations, so that proposing a new type of technical change having the structure developed below can represent nothing more than the choice of a new pair of functions, combined according to the theorem in Section 5. In the same way that Lewbel (1987a, 1987b) has analyzed systems of utility-derived demand, the purpose here is not to invent new types of technical change per se, but rather to "define and delimit" what forms of technical change can be formulated to satisfy the given properties. It is hoped that within this class of technical change transformations are types of technical change that will allow meaningful and fruitful investigations into the nature of technical change and the sources of technological bias.

The next section introduces the notion of an "effective factor price" as a way to account for technical change. Section 3 illustrates that it may be impossible to identify the sources of technological bias although the direction of bias is known. Section 4 provides a group theoretic interpretation of the phenomenon in Section 3 and presents the group properties. Section 5 introduces the additional regularity conditions and fully characterizes the resulting class of transformation groups. Section 6 presents some examples using the characterization theorem and Section 7 summarizes.

2. "EFFECTIVE" FACTOR PRICES

In studies which explicitly incorporate technical change functions, technical change is assumed to affect the productivity of factor inputs so that the true input to the production process is the quantity of "effective" input, that is, the quantity of the input measured in efficiency units. (See, for example, Sato's (1980, p. 769; 1981, p. 22) Definition 1.) To make empirical use of technical change functions, however, requires a different—the dual—interpretation of technical change. This is because econometric models of producer behavior employ duality theory and analyze input demands, and output supply, using price data. For the alternative interpretation of technical change, consider Figure 1.

In Figure 1, MP_i^0 is the period 0 marginal product of, and input-market demand for, an arbitrary factor of production, "factor i," for the firms selling their output in a competitive market at a unit price p. In period 0, the "base period" for subsequent productivity comparisons, suppose that the competitively-determined (real) price of factor i is $(w_i / p)^0$, where w_i is the nominal unit factor price. Then equilibrium in the competitive factor market in period 0 is obtained at A, where the quantity of the input is x_i^0.

Now suppose there is technical progress, or factor i productivity growth, which shifts MP_i^0 to MP_i^1 for period 1. So that we do not complicate the fundamental issues, suppose that the factor supply is perfectly elastic. Then the real price of one unit of factor i remains at $(w_i / p)^0$, factor market equilibrium in period 1 occurs at B, and x_i^1 units of factor i will be employed. An "effective input" approach would interpret the increased employment of the factor as follows: Define

Figure 1

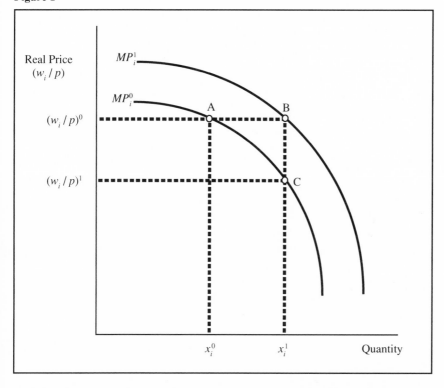

β as $(x_i^1 / x_i^0) - 1$, and define an "efficiency unit of factor i" as being the productive equivalent of 1 unit of factor i in period 0 (the base period). Then roughly speaking, 1 unit of factor i in period 1 can "do the work of" $(1 + \beta)$-units of the "period 0 factor." Alternatively, x_i^0 units of factor i *in period 1* contain $(1 + \beta)x_i^0$ efficiency units. But because the price is given for one unit of the factor, regardless of how many efficiency units it may contain, employment of the more-productive factor must increase in order to drive marginal productivity down to the given level $(w_i / p)^0$. So loosely speaking, the effective *quantity* interpretation of the period 0-to-period 1 employment change is that at a given price *per factor unit*, employment of factor i increases $100\beta\%$ due to the $100\beta\%$ increase in factor i productivity.

For the alternative interpretation, we imagine that one unit of factor i always corresponds to one efficiency unit, and the price of a constant-productivity efficiency unit will be termed the "effective factor price." Then the only way to have an equilibrium quantity of x_i^1 in Figure 1 is for the price of an efficiency unit to have fallen to $(w_i / p)^1$. To actually arrive at an equilibrium at point C in Figure 1 may require a two-step explanation. First, the marginal product curve shifts from MP_i^0 to MP_i^1 to reflect the increased efficiency unit content of a unit of factor i. With a fixed $(w_i / p)^0$ this takes us from A to B in the figure. Now we have to correct for the efficiency unit content of a unit of factor i, and as we do this we must lower the prices or marginal products associated with each quantity of the factor. As we do this, we shift MP_i^1 down and return it to the position of MP_i^0, and the effective price associated with x_i^1 units of the factor falls from $(w_i / p)^0$ to $(w_i / p)^1$. So with fixed productivity of one efficiency unit, the effective price interpretation views the employment change as the response to a simple reduction in the factor cost per efficiency unit, interpreted as the result of a downward shift of the factor supply curve. (In the effective quantity view, the factor supply curve does *not* shift as a result of productivity change, but rather the demand curve shifts, as described above.)

To reconcile the alternative interpretations, we can imagine that w_i / p, the real price of one unit of factor i, is the product of two components: the real price of one *efficiency* unit, and the number of efficiency units per unit of factor i. In both the effective quantity and effective price interpretations, the real price per efficiency unit falls from $(w_i / p)^0$ in period 0 to $(w_i / p)^1$ in period 1, causing a slide down MP_i^0. In the effective quantity view, the number of efficiency units per unit of the factor rises and this shifts MP_i^0 up to MP_i^1; i.e., the fall in the effective price is exactly matched by a rise in the efficiency unit content of a unit of the factor. In the effective price view, the number of efficiency units per unit of the factor does *not* change; hence we remain at point C in the figure.

In summary, treating technical change as a process which alters "effective prices" is similar to assuming that it alters the efficiency unit content of units of factor inputs. The effective price approach, however, permits the introduction of technical change functions into standard econometric models of producer behavior. In this context, the technical change functions will give *effective* factor prices as functions of *observable* factor prices and a parameter representing technical change. In the next section we consider a standard econometric model, introduce technical change functions of a specific form, and show that Stevenson was correct: It is easier to determine the direction of technical bias than to identify its source(s).

3. A CONVENTIONAL ECONOMETRIC APPROACH

Consider a single-output framework under constant returns to scale. A unit output-price function, $p(w, t)$, is given a translog structure in positive factor prices $w_1, w_2, \ldots, w_n$ and time t,

$$
\ln p(w, t) = \alpha_0 + \sum_i \alpha_i \ln w_i + \alpha_T t + 1/2 \sum_i \sum_j \beta_{ij} \ln w_i \ln w_j \\
+ \sum_i \beta_{iT} t \ln w_i + 1/2 \beta_{TT} t^2, \tag{1}
$$

where the parameters satisfy the usual conditions for symmetry, concavity and homogeneity. (See, e.g., Jorgenson, Gollop, and Fraumeni, 1987.) From an assumption of competitive input markets, Shephard's Lemma is employed to derive the cost share (π_i) equations:

$$
\pi_i(w, t) = \frac{\partial \ln p}{\partial \ln w_i} = \alpha_i + \sum_j \beta_{ij} \ln w_j + \beta_{iT} t, \quad i = 1, 2, \ldots, n.
$$

Letting π_T denote the rate of technical change, from (1) we derive the additional equation

$$
-\pi_T(w, t) = \frac{\partial \ln p}{\partial \ln w_i} = \alpha_T + \sum_i \beta_{iT} \ln w_i + \beta_{TT} t.
$$

This indicates that the negative of the rate of technical change is given by the growth rate of the unit output price when all factor prices remain fixed. In the translog model, the biases of technical change are measured by the constant parameters $\beta_{iT} = \partial \pi_i / \partial t$ $(i = 1, 2, ..., n)$. From joint estimation of an n-equation system including $n-1$ of the cost shares and the equation for $-\pi_T$, one concludes that technical change is factor i-using, -neutral, or -saving as the estimate of β_{iT} is positive, zero, or negative, respectively. (See Jorgenson, et al., 1987.)

This approach is enlightening as far as it goes, but suppose technical change actually affects the productivity of each input. This alters the effective price of each factor so that productivity growth (decline) is captured by a fall (rise) in the appropriate factor's effective unit price. As a simple example, consider the case in which all effective factor prices change at a constant exponential rate over time: if w_i' is the effective unit price of factor i, then $w_i' = e^{a_i t} w_i$, where $a_i \neq 0$ is a real constant and w_i is the observed unit factor price ($i = 1$, $2, ..., n$). We would term this "price-augmenting" technical change, the dual analog of factor-augmenting technical change. If we replace w_i in (1) with $w_i' = e^{a_i t} w_i$ and rearrange, we get

$$
\begin{aligned}
\ln p\,(w,\,t) = {} & \alpha_0 + \Sigma_i\,\alpha_i \ln w_i + (\alpha_T + \Sigma_i\,\alpha_i\,a_i)\,t \\
& + 1/2\,\Sigma_i\,\Sigma_j\,\beta_{ij} \ln w_i \ \ln w_j \\
& + \Sigma_i\,(\beta_{iT} + \Sigma_j\,\beta_{ij}\,a_j)\,t \ln w_i \\
& + 1/2(\beta_{TT} + 2\,\Sigma_i\,\beta_{iT}\,a_i + \Sigma_i\,\Sigma_j\,\beta_{ij}\,a_i\,a_j)\,t^2,
\end{aligned}
\tag{2}
$$

from which the biases of technical change are found to be $\partial \pi_i / \partial t = \partial^2 \ln p / \partial t\, \partial \ln w_i = \beta_{iT} + \Sigma_j\,\beta_{ij} a_j$ ($i = 1, 2, ..., n$). That is, the biases of technical change are given by β_{iT} *plus* a weighted sum of the constant growth rates of the effective factor prices (a_j), where the weights are the β_{ij}s. If technical change is factor i-saving, say, the declining share π_i *can* result from declining productivity of factor i ($a_i > 0$, i.e., a rising effective price w_i') causing the term $\beta_{ii} a_i$ to be negative (since β_{ii} must be nonpositive for concavity of p in w). However, factor i's share will be declining even with *improving* productivity ($a_i < 0$) if $\beta_{iT} + \Sigma_{j\pi i} \beta_{ij} a_j$ is negative and greater in absolute value than the nonnegative term $\beta_{ii} a_i$. Hence from a policy-making perspective, the cause of a particular technological bias is of great importance, for a factor's declining share does not necessarily indicate declining productivity of the factor.

While the example here is quite simple, it also convincingly supports Stevenson's observation that the direction of bias is easier to determine than to explain, for although the signs of $\partial \pi_i / \partial t$ ($i = 1, 2, \ldots, n$) can be determined, separate estimation of all of the a_is in (2) is not possible.

4. GROUP THEORY AND TECHNICAL CHANGE

With the benefit of Sato's (1980, 1981) systematic introduction of the theory of Lie transformation groups to economics, it is now easier to understand *why* there have been so few investigations into the nature of technical change. The separation of technical change effects from the effects of scale is a major theoretical problem, and Sato introduced the notion of *holotheticity* as a way to describe this "identifiability problem." The holotheticity of a technology under a particular type of technical change is the situation in which the effect of technical change, operating on the productivities of the inputs, is simply a relabeling of the isoquant map or of the price-space contours of the cost function. In the terminology of group theory, the underlying technology, represented by the production or cost function, is *invariant* under the transformation represented by technical change. In such a situation technical change effects are completely indistinguishable from the effects of scale because they are exactly the same as a scale effect. In the econometric model of Section 3, holotheticity will result if all of the a_is are equal: $a_1 = a_2 = \ldots = a_n = \hat{a}$. Then $\ln p$ collapses to

$$\ln p(w, t) = \alpha_0 + \Sigma_i \alpha_i \ln w_i + (\alpha_T + \hat{a}) t$$

$$+ 1/2 \Sigma_i \Sigma_j \beta_{ij} \ln w_i \ln w_j \qquad (3)$$

$$+ \Sigma_i \beta_{iT} t \ln w_i + 1/2\beta_{TT} t^2,$$

i.e., the form of $\ln p$ in (1) plus $\hat{a} t$, since $\hat{a}$ can be moved to the outside of the summations and $\Sigma_i \alpha_i = 1$ and $\Sigma_i \beta_{ij} = \Sigma_j \beta_{ij} = \Sigma_i \beta_{iT} = 0$ from the homogeneity of p in w. Hence the technical change effect is indistinguishable from a scale factor $e^{\hat{a}t}$ in the price function $p(w, t)$.

This identifiability problem is not a recently recognized phenomenon; it was known over a quarter century ago to Solow (1961, p. 67): "The problem of measuring economies of scale and distinguishing their effects from those of technical progress is an econometric puzzle worthy of anybody's talents." One of Sato's contributions toward solving this puzzle is in formalizing a theoretical model of exogenous technical change in which we can clearly see the conditions present when scale effects and the effects of technical change are indistinguishable. In addition, it is apparent from the first pages of Sato (1980 or 1981) that holotheticity *has* been a problem historically because researchers have primarily used only factor-augmenting types of technical change, and the factor-augmenting type is a "bad" hypothesis if scale effects are to be distinguished from technical change effects when the production function is homogeneous, as is often assumed.[1]

In recent years empirical researchers have utilized duality theory and cost functions to analyze producer behavior and technical change. In this context "price-augmenting" technical change is a poor hypothesis because of the homogeneity of the rational firm's cost function in factor prices; this can be seen in Section 3 from the impossibility of estimating the a_is in (2). Then it is evident that another contribution made by Sato (1981) is the presentation of new types of technical change functions which will *not* lead to the extreme identifiability problem which the factor- and price-augmenting types produce. The structure given to these new types is the structure of Lie groups of transformations.

While Sato's analysis was set in the context of a neoclassical production function, empirical analyses of producer behavior and technical change generally use a cost function approach. Therefore, consider the following analysis to be based on a model of production and technical change in which n inputs ($n \geq 2$) are employed for the production of m outputs ($m \geq 1$). (Since empirical implementation will be based on a cost function, there is no need to restrict analysis to the single-output case.) Let the inputs be represented by the real column vector $x = (x_1, x_2, ..., x_n)^T \geq 0$, and let their associated unit prices be represented by the real vector $w = (w_1, w_2, ..., w_n)^T \geq 0$.[2] Productivity change is assumed to occur exogenously, have no effect on the mathematical form of the relevant cost function, and enter the process of production by altering the productivities of the factors of production as reflected by their associated effective unit prices. In the production function approach (Sato, 1980, 1981) it would be assumed that there exists a well-behaved, real-valued function which defines the transformation of a "raw," or "nominal," factor, x_i, into the "effective" factor, x_i' ($i = 1, 2, ..., n$). In the dual approach we assume the existence of a continuous, nonnegative, real-valued function which defines the transformation of a "nominal" price, w_i, into an "effective" price, w_i' ($i = 1, 2, ..., n$). Ongoing technical change is driven by changes in a real-valued "technical change parameter." In nearly all, if not all, empirical studies, "time" has been the lone technical change parameter, or the "technology variable," so the passage of time is assumed to drive exogenous technical change. Then the transformation of nominal input prices into effective (unobservable) input prices is defined by the technical change functions $\phi_i(w, t)$ ($i = 1, 2, ..., n$), where $w \geq 0$ and $t \in \mathbb{R}$:

$$w' = (w_1', w_2', ..., w_n')^T = [\phi_1(w, t), \phi_2(w, t), ..., \phi_n(w, t)^T = \phi(w, t),$$

where t represents time. We will assume that $\phi(w, t) \geq 0$ holds for all $w \geq 0$ and for all t. (The strict positivity of at least one factor price has already been imposed. If $w_k > 0$ it seems reasonable to require $\phi_k(w, t) > 0$; permitting $\phi_k(w, t) = 0$ with $w_k > 0$ allows factor k to become "infinitely productive," a situation unlikely to be found, however desirable.) The functions f_i may be differentiable, but differentiability is *not* assumed in any of the following analysis; *only continuity is presumed.*

Sato's results are based on the theory of continuous (Lie) groups of transformations, so suppose that ϕ satisfies the group properties as well.

Property 1 ("composition" property). For all feasible real numbers t_1 and t_2 which define successive effective input price vectors $w' = \phi(w, t_1)$ and $w'' = \phi(w', t_2) = \phi[\phi(w, t_1), t_2]$, there exists a real number t_3 depending only on t_1 and t_2, $t_3 = G(t_1, t_2)$, such that the equation of alternative expressions for w'',

$$\phi[\phi(w, t_1), t_2] = \phi[w, G(t_1, t_2)], \qquad (4)$$

is an identity in the w_is $(i = 1, 2, \ldots, n)$, t_1 and t_2.

Property 2 (existence of an inverse transformation). For every feasible real value t, there exists a real value τ such that the technical change "caused" by t can be "undone" by τ: $w'' = \phi(w', \tau) = \phi''[\phi(w, t), \tau] = w$, for all $w \geq 0$.

Technical change functions, $\phi_i(w, t)$ $(i = 1, 2, \ldots, n)$, satisfying the above properties constitute a *one-parameter group of transformations.*[3] Sato (1981, p. 27) noted that "it may be possible to construct a type of technical progress which is meaningful, yet does not satisfy the group properties. However, it should be noted that all of the known types discussed in the economic literature thus far do in fact satisfy the [group] assumptions." At least partially for this reason, investigating the functional forms of technical change functions through the group properties seems quite reasonable.

5. A SUBCLASS OF ONE-PARAMETER GROUPS
OF TRANSFORMATIONS

In order to extend the currently limited list of one-parameter groups of technical change transformations from Mitchell (1987), suppose f satisfies two additional conditions. First, "weak left reducibility":

$$\text{For every } w \geq 0,\ \phi(w, t_1) = \phi(w, t_2) \Rightarrow t_1 = t_2.^4 \qquad (5)$$

This condition means that to get a particular effective input price vector from given nominal input prices (and an unchanging process of exogenous technical change represented by ϕ) there is a unique length of time to wait.[5] While the continuity of f and (5) imply strict monotonicity of $\phi_i(w, t)$ in t ($i = 1, 2, \ldots, n$), the popular factor- and price-augmenting technical change functions, $\phi_i(y, t) = e^{a_i t} y_i$ with a_i a real constant ($i = 1, 2, \ldots, n$), satisfy (5).

Secondly, suppose w_1 is fixed at w_1^0.[6] Then write

$$w' = \phi \left[\begin{array}{c} 0 \\ w_1 \\ w_2 \\ \vdots \\ w_n \end{array}, t \right] = \chi \left[\begin{array}{c} t \\ w_2 \\ w_3 \\ \vdots \\ w_n \end{array} \right] = \chi(y) \geq 0, \qquad (6)$$

so that (6) defines a function χ whose argument is a vector variable $y \in \mathbb{R} \times \mathbb{R}^{n-1}_+$. The additional condition on ϕ is

$$\text{There exists a } w_1^0 \text{ so that } \chi(y) = z$$
$$\text{can be solved uniquely for } y = \psi(z),\ z \geq 0. \qquad (7)$$

As y is "partitioned," let the vector-valued function y be partitioned as $y(z) = \left[\begin{smallmatrix} H(z) \\ h(z) \end{smallmatrix} \right]$, so that $H(z)$ is scalar-valued but $h(z) \in \mathbb{R}^{n-1}$ is vector-valued, for $z \geq 0$.[7] In particular, note that the first component of ψ, $H(z)$, gives the length of time required for the nominal price w_1^0 to become an effective price w_1'. Being able to compute such a t may be useful in practice.

With $\chi[(t, w_2, \ldots, w_n)^T]$ uniquely invertible and $\phi(w, t)$ weakly left reducible, we focus on the composition property, (4), and identify the general form of transformation schemes satisfying the three conditions (4), (5), and (7). The solutions are given by the following:

Lemma 1 *(Aczél, 1966). If there exists a w_1^0 so that $\chi(y) = \chi[(t, w_2, \ldots, w_n)^T] = z$ can be solved uniquely for y, where χ is defined by (6), and ϕ is weakly left reducible, then*

$$\phi(w, t) = \psi^{-1}\left[\begin{array}{c} G[H(w), t] \\ h(w) \end{array}\right], \tag{8}$$

with arbitrary $\psi(w) = [\begin{smallmatrix} H(w) \\ h(w) \end{smallmatrix}]$ having a unique inverse, is the general solution of (4), the so-called "transformation equation."

Proof. *Aczél (1966, pp. 364-366).*

Lemma 1, by itself, does not necessarily yield a group, for we have not guaranteed that Property 2 will be satisfied. It is G that must be of such a form that inverse (and identity) transformations exist. However, the weak left reducibility of ϕ and Property 1 impart some structure to G, namely "associativity."

Suppose productivity change occurs for three separate periods of time during which the technical change parameter takes the respective values t_1, t_2, and t_3. At the end of the first period the effective factor prices are given by $w' = \phi(w, t_1)$, at the end of the second period by $w'' = \phi(w', t_2)$, and at the end of the third period by $w''' = \phi(w'', t_3)$. Then by substitution, w''' can be expressed in the following ways:

$$w''' = \phi[\phi(w', t_2), t_3], \tag{9}$$

$$= \phi\{\phi[\phi(w, t_1), t_2], t_3\}. \tag{10}$$

From (4), $\phi[\phi(w, t_1), t_2] = \phi[w, G(t_1, t_2)]$. Then from (10), $w''' = \phi$ $\{\phi[w, G(t_1, t_2)], t_3\}$ and applying (4) to this form yields

$$w''' = \phi\{w, G[G(t_1, t_2), t_3]\}. \tag{11}$$

Also from (4), (9) can be rewritten as $w''' = \phi[w', G(t_2, t_3)]$, but $w' \equiv \phi(w, t_1)$, so $w''' = \phi[\phi(w, t_1), G(t_2, t_3)]$. Applying (4) again yields

$$w''' = \phi\{w, G[t_1, G(t_2, t_3)]\}. \tag{12}$$

From (11), (12) and the weak left reducibility of ϕ, the associativity of G is found:

$$G[G(t_1, t_2), t_3] = G[t_1, G(t_2, t_3)]. \tag{13}$$

Along with the associativity of G, another result we will find useful later is:

Lemma 2. *If $\phi(w, t)$ is weakly left reducible and satisfies the group properties, then $G(t, t_0) = G(t_0, t) = t$, for all feasible t, where $t_0 = G(t, t) \in \mathbb{R}$ gives the identity transformation and τ determines the particular inverse transformation for an arbitrarily chosen t.*

Proof. By Property 1, $\phi[\phi(w, t), \tilde{t}] = \phi[w, G(t, \tilde{t})]$ for all feasible t and $\tilde{t}$. Allow technical change to occur for a third period during which the technical change parameter takes the value $\bar{t}$. Then we have

$$\phi\{\phi[\phi(w, t), \tilde{t}], \bar{t}\} = \phi\{\phi[w, G(t, \tilde{t})], \bar{t}\}, \quad \text{for all } t, \tilde{t}, \bar{t}. \tag{14}$$

Using Property 2, let $\tilde{t} = \tau$ be the value of the parameter which "undoes" the technical change done by t: $\phi[\phi(w, t), \tau] = w$ for all $w \geq 0$. Then the left side of (14) is $\phi(w, \bar{t})$. Let the real value $G(t, \tau)$ be denoted by t_0; this defines the identity transformation since $\phi[w, G(t, \tau)] = w = \phi[\phi(w, t), \tau]$ for all $w \geq 0$. Then the right side of

(14) is $\phi\,[\phi(w,\,t_0),\,\overline{t}\,)]$, which by (4) is equivalent to $\phi\,[w,\,\,G(t_0,\overline{t}\,)]$. Equating the left side of (14) to the right side gives $\phi\,(w,\,\overline{t}\,) = \phi[w,\,G(t_0,\overline{t}\,)]$, which yields $\overline{t} = G(t_0,\overline{t}\,)$ for all $\overline{t}$ by the weak left reducibility of ϕ.

To prove that $G(t,\,t_0) = t$ for all t, start with (14) and apply Property 2: Let $\overline{t}$ be the value of the parameter which "undoes" the technical change done by $\tilde{t}$, so on the left side of (14) we have $\phi\{\phi[\phi(w,\,t),\tilde{t},\,t\,],\,t\,\} = \phi(w,\,t)$ for all $\phi(w,\,t) \geq 0$. By Property 1 the right side of (14) can be expressed as $\phi\{w,\,G[G(\tilde{t},\,t\,),\,\overline{t}]\}$, and by the associativity of G this can be written as $\phi\{w,\,G[t,\,G(\tilde{t},\overline{t}\,)]\}$. Now let the real value $G(\tilde{t},\overline{t}\,)$ be denoted by t_0. (Again this defines the identity transformation since $\phi[\phi(w,\,t\,),\,G(\tilde{t},t\,)] = \phi(w,\,t\,) = \phi\{\phi[\phi(w,\,t),\tilde{t}\,],\,\overline{t}\,\}$ for all $\phi(w,\,t) \geq 0$.) Equating the left side of (14) to the right side gives $\phi(w,\,t) = \phi[w,\,G(t,\,t_0)]$ for all t, which yields $t = G(t,\,t_0)$ by the weak left reducibility of ϕ.

Q.E.D.

Keeping in mind that the objective of the paper is to enable researchers to generate new functional forms for transformation groups rather than a complete characterization of all one-parameter groups, suppose that G is "reducible" on both sides:

$$G(t,\,t_1) = G(t,\,t_2) \text{ for all } t \Rightarrow t_1 = t_2, \text{ and} \tag{15}$$

$$G(t_1,\,t) = G(t_2,\,t) \text{ for all } t \Rightarrow t_1 = t_2.$$

It is clear that G is already continuous in both arguments, so (15) imposes strict monotonicity in both arguments.[8] Now the choice of a function G can be simplified to choosing an arbitrary, continuous and strictly monotonic function with a single (scalar) argument.

Lemma 3 *(Aczél, 1966). If with t_1 and t_2, $G(t_1,\,t_2)$ is always in a given interval (possibly infinite), and (15) holds, then*

$$G(t_1,\,t_2) = \Pi[\Gamma^{-1}(t_1) + \Gamma^{-1}(t_2)], \tag{16}$$

with continuous and strictly monotonic Γ, is the general continuous solution of the "associativity equation," (13).

Proof. *(Aczél, 1966, pp. 256-267).*

It is clear from (16) that in addition to being "quasilinear" (Aczél, 1966, p. 151), G is symmetric: $G(t_1, t_2) = G(t_2, t_1)$ for all t_1, t_2. Under the conditions of Lemmas 2 and 3 it is also apparent that G must satisfy $\Gamma[\Gamma^{-1}(t) + \Gamma^{-1}(t_0)] = t$ for all t, i.e.

$$\Gamma^{-1}(t) + \Gamma^{-1}(t_0) = \Gamma^{-1}(t), \quad \text{for all } t. \tag{17}$$

If $\Gamma^{-1}(t)$ is defined for all t, then (17) implies $\Gamma^{-1}(t_0) = 0 \Leftrightarrow t_0 = \Gamma(0)$.[9] Since $t_0 = G(t, \tau)$, where τ defines the inverse transformation of $\phi(w, t)$, Γ must also satisfy

$$\Gamma[\Gamma^{-1}(\tau) + \Gamma^{-1}(\tau)] = t_0 \Leftrightarrow \Gamma^{-1}(t) + \Gamma^{-1}(\tau) = \Gamma^{-1}(t_0). \tag{18}$$

If $\Gamma^{-1}(t)$ is defined for all t, $\Gamma^{-1}(t_0) = 0$ and $\Gamma^{-1}(\mathrm{t}) = \Gamma^{-1}(t) \Leftrightarrow \tau = \Gamma[-\Gamma^{-1}(t)]$.[10] So the choice of G has been simplified to choosing a continuous and strictly monotonic function Γ, defined at $-\Gamma^{-1}(t)$ for all t and the origin.

Returning to Lemma 1, rewrite (8) using (16) to get

$$\phi(w, t) = \psi^{-1} \left[\begin{array}{c} \Gamma\{\Gamma^{-1}[H(w)] + \Gamma^{-1}(t)\} \\ h(w) \end{array} \right] \tag{19}$$

as a transformation ϕ which will possess the properties of a transformation group and satisfy (5), (7) and (15) from choices of a continuous and strictly monotonic Γ, and $\psi(w) = [\frac{H(w)}{h(w)}]$ with unique inverse. From (19) it is apparent that Γ^{-1} must be defined for all possible values of H, i.e., the range of H must be a subset of the range of Γ.[11] By inspection it is apparent that $\Gamma(0)$ yields the identity transformation and it is a simple matter to use (19) to show that $\Gamma[-\Gamma^{-1}(t)]$ is the value of the parameter which gives the inverse transformation for $\phi(w, t)$ for all feasible t. Then what has been proved in this section (with trivial aspects of the proof of sufficiency omitted) is the following:

Theorem. *Let $\phi(w, t) \geq 0$ be a technical change transformation for $w \geq 0$ and real t. Then ϕ is weakly left reducible, satisfies (7) and the group properties with G reducible on both sides **if and only if** ϕ is of the form in (19) for an arbitrary $\psi(z) = [^{H(w)}_{h(w)}]$ with a unique inverse and an arbitrary continuous and strictly monotonic function Γ defined at the origin and $-\Gamma^{-1}(t)$ for all possible t, and with a range containing the range of H.*

6. ILLUSTRATIVE EXAMPLES

Example 1. Although its shortcomings in empirical models has been noted, the price-augmenting type in its familiar exponential form is easily generated using (19). When the technical change parameter is "time," the most appealing form for G is additive: $G(t_1, t_2) = t_1 + t_2$, $t_1, t_2 \in \mathbb{R}$. This form is derived from $\Gamma(t) = t$, so $t_0 = 0$ defines the identity transformation, and for any real t, $\tau = -t$ defines the inverse transformation. (The interpretation of the additive form of G is that waiting t_1 years then t_2 years will yield the same effective price vector as waiting $t_1 + t_2$ years from the outset.) Let $\psi(z) = [^{H(z)}_{h(z)}]$ be formed from

$$H(z) = \ln z_1^{1/a_1}, \; h(z) = \begin{bmatrix} z_1^{-a_2/a_1} z_2 \\ z_1^{-a_3/a_1} z_3 \\ \vdots \\ z_1^{-a_n/a_1} z_n \end{bmatrix}, z_1 > 0, \qquad (20)$$

where $a_i \neq 0$ is a real constant ($i = 1, 2, ..., n$). Solving for ψ^{-1} yields

$$\psi^{-1}(z) = \begin{bmatrix} \exp(a_1 z_1) \\ z_2 \exp(a_2 z_1) \\ \vdots \\ z_n \exp(a_n z_1) \end{bmatrix},$$

so that the technical change functions defined by (19) are

$$\phi(w, t) = \psi^{-1}\left[\begin{pmatrix} \Gamma\{\Gamma^{-1}[H(w)] + \Gamma^{-1}(t)\} \\ h(w) \end{pmatrix}\right]$$

$$= \psi^{-1}\left[\begin{pmatrix} H(w) + t \\ h(w) \end{pmatrix}\right] = \begin{bmatrix} e^{a_1 t} w_1 \\ e^{a_2 t} w_2 \\ \vdots \\ e^{a_n t} w_n \end{bmatrix}.$$

Example 2. One characteristic of some of the technical change functions introduced by Sato (1981) which is new is *additivity*. Empirical studies have generally, if not always, considered multiplicative (i.e., augmenting) forms of technical change. But consider $\Gamma(t) = t$ ($t \in \mathbb{R}$) again, and let

$$\psi(z) = \begin{bmatrix} H(z) \\ h(z) \end{bmatrix} = \begin{bmatrix} z_1 / a_1 \\ z_2 - (a_2 / a_1) z_1 \\ \vdots \\ z_n - (a_n / a_1) z_1 \end{bmatrix}, \tag{21}$$

where a_i is a nonzero real constant ($i = 1, 2, \ldots, n$). Solving for $\psi^{-1} = \chi$ yields

$$\psi^{-1}(z) = \begin{bmatrix} a_1 z_1 \\ z_2 + a_2 z_1 \\ \vdots \\ z_n + a_n z_1 \end{bmatrix},$$

from which the technical change functions are found:

$$\phi(w, t) \;=\; \psi^{-1}\!\begin{bmatrix} H(w) + t \\ h(w) \end{bmatrix} = \begin{bmatrix} w_1 + a_1 t \\ w_1 + a_2 t \\ \vdots \\ w_n + a_n t \end{bmatrix}.$$

This is the *additive type* of technical change (Sato, 1981) under which effective prices change at a constant rate: $\partial \phi_i / \partial t = a_i$, since ϕ_i is partially differentiable ($i = 1, 2, \ldots, n$).

Example 3. Of course one can combine one type of technical change for one factor with technical change of another type for a different factor. Let $n = 4$, $\Gamma(t) = t$ ($t \in \mathbb{R}$) and

$$\psi(z) = \begin{bmatrix} H(z) \\ h(z) \end{bmatrix} = \begin{bmatrix} \ln z_1^{\,1/a_1} \\ z_2 - (a_2 / a_1) \ln z_1 \\ z_3 - (a_3 / a_1) z_4 \ln z_1 \\ z_4 \end{bmatrix}, \; z_1 > 0, \quad (22)$$

with a_i a nonzero real constant ($i = 1, 2, 3$). Then

$$\psi^{-1}(z) = \begin{bmatrix} \exp(a_1 z_1) \\ z_2 + (a_2 z_1) \\ z_3 + a_3 z_1 z_4 \\ z_4 \end{bmatrix},$$

from which we find

$$\phi(w, t) \;=\; \psi^{-1}\!\begin{bmatrix} H(w) + t \\ h(w) \end{bmatrix} = \begin{bmatrix} e^{a_1 t} w_1 \\ w_2 + a_2 t \\ w_3 + a_3 w_4 t \\ w_4 \end{bmatrix}.$$

Here technical change is of the (exponential) price-augmenting type for the first price; the additive type for the second; the "ratio additive" type for the third (see Sato, 1981); and there is *no* productivity change for the fourth factor. Now it can be seen that there are many different types of transformations, $\phi(w, t)$, if one combines the basic types— augmenting, additive, ratio additive, "none," etc.—in various ways through the price vector w.

Example 4. Consider a different form for Γ: $\Gamma(t) = e^t$, $t \in \mathbb{R}$. Then $\Gamma^{-1}(t) = \ln t$, $t > 0$, and $G(t_1, t_2) = t_1 t_2$. $t_0 = 1$ gives the identity transformation and $\tau = 1/t$ gives the inverse transformation of $\phi(w, t)$ for any $t > 0$. (The technical change parameter t is no longer sensibly interpreted as "time," but rather as some kind of index number whose exogenous changes generate technical change.) Now let $\psi(z) = [\frac{H(z)}{h(z)}]$ be formed from $H(z) = z_1^{1/a_1}$, $z_1 > 0$, and the $h(z)$ in (20). Then finding ψ^{-1} yields

$$\psi^{-1}(z) = \begin{bmatrix} z_1^{a_1} \\ z_1^{a_2} z_2 \\ \vdots \\ z_1^{a_n} z_n \end{bmatrix}$$

from which we find the technical change functions,

$$\phi(w, t) = \psi^{-1}\left[\begin{pmatrix} \Gamma\{\Gamma^{-1}[H(w)] + \Gamma^{-1}(t)\} \\ h(w) \end{pmatrix}\right]$$

$$= \psi^{-1}\left[\begin{pmatrix} tH(w) \\ h(w) \end{pmatrix}\right] = \begin{bmatrix} t^{a_1} w_1 \\ t^{a_2} w_2 \\ \vdots \\ t^{a_n} w_n \end{bmatrix}.$$

Example 5. Consider another form for G: from $\Gamma(t) = e^t - 1$ $(t \in \mathbb{R})$, $G(t_1, t_2) = t_1 + t_1 t_2 + t_2$, where $t_1, t_2 > -1$. For this G, $t_0 = 0$ and $\tau = -t \, / \, (1 + t)$ give the identity and inverse transformations, respectively. Now let $\psi(z)$ be formed from $H(x) = z_1^{\,1/a_1} - 1, z_1 > 0$, and the $h(z)$ from (20) again. Finding ψ^{-1} yields:

$$\psi^{-1}(z) = \begin{bmatrix} (z_1 + 1)^{a_1} \\ (z_1 + 1)^{a_2} z_2 \\ \vdots \\ (z_1 + 1)^{a_n} z_n \end{bmatrix}$$

Then the technical change functions are given by

$$\phi(w, t) = \psi^{-1}\left[\begin{pmatrix} \Gamma\{\Gamma^{-1}[H(w)] + \Gamma^{-1}(t)\} \\ h(w) \end{pmatrix} \right]$$

$$= \psi^{-1}\left[\begin{pmatrix} H(w) + tH(w) + t \\ h(w) \end{pmatrix} \right] = \begin{bmatrix} (1 + t) & w_1 \\ (1 + t) & w_2 \\ \vdots \\ (1 + t) & w_n \end{bmatrix}^{\begin{array}{c}a_1 \\ a_2 \\ \\ a_n\end{array}}, \; t > -1.$$

While the form of G eludes intuitive interpretation, if t is time these (price-augmenting) technical change functions have the property that their rates of growth decline over time in absolute value (in contrast to the exponential price-augmenting type which exhibits a constant growth rate):

$$\frac{\partial \ln \phi_i}{\partial t} = \frac{a_i}{1 + t} \qquad (i = 1, 2, \ldots, n).^{[13]}$$

7. SUMMARY

This paper calls attention to the dearth of econometrically identifiable functional forms for technical change functions and has developed the means with which a researcher can generate new forms of technical change functions satisfying the appealing properties of transformation groups. By assuming strict monotonicity of the technical change functions in the exogenous variable "causing" technical change (possibly by proxy), the unique invertibility of χ defined by (6), and strict monotonicity in both arguments of the function G in (4), we were able to completely characterize the resulting subclass of one-parameter groups of [technical change] transformations. It is hoped that the results here will lead to more rigorous, and useful, empirical studies of producer behavior under technical change.

NOTES

1. See Sato (1980, 1981) on the "Solow-Stigler controversy."
2. If $\mathbb{R}^n_+$ is the set of all real, nonnegative, n-dimensional vectors, then $w \geq 0$ means $w \in \mathbb{R}^n_+$ but $w \neq 0$, where the 0 represents a column vector of n zeroes. It seems unnecessary to worry about $w = 0$; we will presume at least one factor is strictly positive.
3. A third property is the existence of an identity transformation: There exists a real constant, t_0, such that for all $w \geq 0$ no technical change occurs, $w' = \phi(w, t_0) = w$. However, given Properties 1 and 2, clearly the value of t_0 is $G(t, \tau)$. In addition, a fourth property, associativity, is omitted because it "holds trivially" in the present context (Sato, 1981, p. 26; Mitchell and Primont, 1988). Finally we will refer to these groups of transformations "generically"; to call them "Lie groups" would not be correct because Lie considered analytical (i.e., differentiable) transformations and we have not imposed differentiability.

4. See Aczél (1966). An example of a technical change transformation which is not weakly left reducible for all $w \geq 0$ is $\phi(w, t) = \begin{pmatrix} w_1^t \\ w_2^t \end{pmatrix}$ for $t > 0$. Here $G(t_1, t_2) = t_1 t_2$ and $\tau = 1/t$ satisfy Properties 1 and 2, but for $\tilde{w} = \begin{pmatrix} 1 \\ 1 \end{pmatrix}$, $\phi(\tilde{w}, t_1) = \phi(\tilde{w}, t_2)$ for any positive values t_1 and t_2.

5. This of course presumes that the particular w' is feasible given current nominal prices, w. That is, w' is in the image set of $\phi(w, t)$.

6. Since the vectors w and w' can be rearranged, the choice of w_1 is completely arbitrary.

7. $\chi(y)$ and $\psi(z)$ are each other's inverses: $\chi(y) = \psi^{-1}(y)$. Hence $H[\psi^{-1}(y)] = y_1 = t$ and $h[\psi^{-1}(y)] = (y_2, \ldots, y_n)^T = (w_2, \ldots, w_n)^T$.

8. More than simply "strictly monotonic" in both arguments, Aczél (1966, pp. 255-256) shows that for G continuous, associative and reducible on both sides, G is actually strictly *increasing* in both arguments.

9. If Γ is not defined at the origin, then there is no constant t_0 giving the identity transformation for all $w \geq 0$. As an example consider $\Gamma(t) = \ln t$ $(t > 0)$ and $\psi(z) = \begin{pmatrix} \ln z_1 \\ z_2 \end{pmatrix}$, $z_1 > 0$. Then $G(t_1, t_2) = \ln(e^{t_1} + e^{t_2})$ and from (8), $\phi(w, t) = \begin{pmatrix} w_1 + e^t \\ w_2 \end{pmatrix}$. $\phi(w, t_0) = w$ for all $w \geq 0$ and $G(t, t_0) = t$ for all t only in the limit as $t_0 \rightarrow -\infty$, hence there is no constant $t_0 \in \mathbb{R}$ such that $\phi(w, t_0) = w$ for all w. (Note that Property 2 is violated).

10. If Γ were not defined at $-\Gamma^{-1}(t)$ for all possible values of t, then there will not be a τ for every possible t satisfying Property 2 for all $w \geq 0$. As an example, consider $\Gamma(t) = t^{1/2}$ $(t \geq 0)$ and $\psi(z) = \begin{pmatrix} z_1^{1/2} \\ z_2 \end{pmatrix}$. Then $G(t_1, t_2) = (t_1^2 + t_2^2)^{1/2}$ and from (8), $\phi(w, t) = \begin{pmatrix} w_1 + t^2 \\ w_2 \end{pmatrix}$. For there to exist an inverse transformation for all possible values of t, we must have a τ solving $t^2 + \tau^2 = 0 = t_0 = \Gamma(0)$ for all t. Clearly if $t \neq 0$, there is no $\tau \in \mathbb{R}$ satisfying Property 2.

11. Consider $\Gamma(t) = e^t \Leftrightarrow \Gamma^{-1}(t) = \ln t$ $(t > 0)$ and suppose $H(w) = \ell n w_1$ $(w_1 > 0)$. (These are the forms implied by the group in Note 4 which is not weakly left reducible for all w). Then, $\Gamma^{-1}[H(w)] = \ln (\ln w_1)$ which is not defined for $w_1 \leq 1$.

12. Here it should be pointed out that the domain restriction on t for ϕ applies only to Γ^{-1}; Γ itself is never evaluated at t per se. Hence there is no inconsistency with Γ defined everywhere in $\mathbb{R}$ in this example although ϕ and Γ^{-1} may accept only positive values of t.

13. It is true that Example 2 has this property, but in Example 2 we have the possibility of $w_k > 0$ and $\phi_k(w, t) \leq 0$ which we wish to exclude; Example 5 avoids this complication. Further if these transformations are defined for inputs rather than prices, then the additive form in Example 2 violates the positive input requirement, $x_i = 0 \Rightarrow \phi_i(w, t) = 0$ for all possible t, a property introduced by Mitchell and Primont (1988). Example 5 satisfies this property while exhibiting a declining growth rate over time.

REFERENCES

Aczél, J. D. (1966). *Lectures on functional equations and their applications.* Mathematics in science and engineering, vol. 19. New York: Academic Press.

Baltagi, B. H., and Griffin, J. M. (1988). A general index of technical change. *Journal of Political Economy 96,* 20-41.

Jorgenson, D. W., Gollop, F. M., and Fraumeni, B. M. (1987). *Productivity and U.S. economic growth.* Cambridge, MA: Harvard University Press.

Lewbel, A. (1987a). Characterizing some Gorman Engel curves. *Econometrica 55,* 1451-1459.

Lewbel, A. (1987b). Fractional demand systems. *Journal of Econometrics 36,* 311-337.

Mitchell, T. M. (1987). Toward empirical applications of Lie-group technical progress functions. *Economics Letters 25,* 111-116.

Mitchell, T. M., and Primont, D. (1988). On the group properties of technical change functions. Southern Illinois University, Department of Economics.

Sato, R. (1980). The impact of technical change on the holotheticity of production functions. *Review of Economic Studies 47,* 767-776.

Sato, R. (1981). *Theory of technical change and economic invariance.* New York: Academic Press.

Solow, R. M. (1961). Comment (on Stigler). *Output, input and productivity measurement,* pp. 64-67. Income and Wealth Series. Princeton: Princeton University Press.

Stevenson, R. (1980). Measuring technological bias. *American Economic Review 70,* 162-173.

Index